# Cold and Ultracold Collisions in Quantum Microscopic and Mesoscopic Systems

Cold and ultracold collisions occupy a strategic position at the intersection of several powerful themes of current research in chemical physics; in atomic, molecular, and optical physics; and even in condensed matter. The nature of these collisions has important consequences for optical manipulation of inelastic and reactive processes, precision measurement of molecular and atomic properties, matter–wave coherences, and quantum-statistical condensates of dilute, weakly interacting atoms. This crucial position explains the wide interest and explosive growth of the field since its inception in 1987. The author reviews elements of quantum scattering theory, collisions taking place in the presence of one or more light fields, and collisions in the dark, below the photon recoil limit imposed by the presence of any light field. Finally, it reviews the essential properties of these mesoscopic quantum systems, and describes the key importance of the scattering length to condensate stability.

JOHN WEINER graduated from the Pennsylvania State University with high distinction and received his Ph.D. from the University of Chicago. He is a Fellow of the American Physical Society and the Washington Academy of Sciences. He is a member of numerous professional societies in the USA and in France. He has held lectureships and professorships at Yale University, Dartmouth College, and the University of Maryland in the USA. He is presently a professor of chemical physics at the Université Paul Sabatier in Toulouse. France.

T0254465

# Cold and Ultracold Collisions in Quantum Microscopic and Mesoscopic Systems

JOHN WEINER
Université Paul Sabatier

CAMBRIDGE
UNIVERSITY PRESS

CAMBRIDGE UNIVERSITY PRESS
Cambridge, New York, Melbourne, Madrid, Cape Town, Singapore, São Paulo

Cambridge University Press
The Edinburgh Building, Cambridge CB2 8RU, UK

Published in the United States of America by Cambridge University Press, New York

www.cambridge.org
Information on this title: www.cambridge.org/9780521781213

First published 2003
This digitally printed version 2007

*A catalogue record for this publication is available from the British Library*

*Library of Congress Cataloguing in Publication data*

Weiner, John, 1943–
  Cold and ultracold collisions in quantum microscopic and mesoscopic systems / John Weiner.
    p.  cm.
  Includes bibliographical references and index.
  ISBN 0 521 78121 3
  1. Collisions (Nuclear physics)   2. Materials at low temperatures.   I. Title.
QC794.6.C6W45 2003
539.7´57–dc21   2003044033

ISBN 978-0-521-78121-3 hardback
ISBN 978-0-521-03693-1 paperback

# Contents

# Preface

Cold and ultracold collisions occupy a strategic position at the intersection of several power-ful themes of current research in chemical physics, in atomic, molecular and optical physics, and even in condensed matter. The nature of these collisions has important consequences for optical manipulation of inelastic and reactive processes, precision measurement of molec-ular and atomic properties, matter–wave coherence and quantum-statistical condensates of dilute, weakly interacting atoms. This crucial position explains the wide interest and ex-plosive growth of the field since its inception in 1987. Obviously due to continuing rapid developments the very latest new results cannot appear in book form, but the field is suf-ficiently mature that a fairly comprehensive account of the principal research themes can now be undertaken. The hope is that this account will prove useful to newcomers seeking a point of entry and as a reference for those already initiated.

After a general introduction and a brief review of the elements of scattering theory in Chapters 1 and 2, the next four chapters treat collisions taking place in the presence of one or more light fields. The reason for this is simply historical. After the development of the physics of optical cooling and trapping from the early to mid 1980s, the first generation of collisions experiments applied this light-force physics to cool and confine atoms in traps and beams. The dipole force trap, the magneto-optical trap, and "brightened" atom beams, described in Chapter 3, became the experimentalists' tools of the trade. Light fields from the radiation pressure force and the dipole gradient potential, needed to cool atoms to submillikelvin temperatures, also play an active role in the collision processes themselves. Chapter 4 describes how inelastic, energy-releasing collisions can be detected and studied by losses from atom traps, and Chapter 5 reviews the important advances derived from precision photoassociation spectroscopy. Chapter 6 recounts how light not only directs scattering flux into inelastic and reactive channels, but also prevents close atom encounters from occurring – atoms can be "shielded" from collisions and the scattering flux redirected to elastic channels.

Although cooling and trapping with light fields opened the way to the study submillikelvin collisions, these fields themselves posed a barrier to the lower temperatures and higher densities required to reach the quantum-degenerate regime. Evaporative cooling of ground-state atoms and confinement in dark magnetic traps provided the pathway to breach this light-field barrier. The physics of binary collisions in this regime reduces to s-wave scattering, which is much simpler than the inelastic and reactive processes at higher temperatures, but critically important for the structure and dynamics of Bose–Einstein condensates. Chapter 7 reviews the essential properties of these mesoscopic quantum systems, and describes the key importance of the scattering length to condensate stability. A vital area of current research focuses on the active manipulation of the scattering length by external fields and the search for quantum-degenerate molecular species. This story is far from complete, but enough has already been accomplished to point the reader toward those areas likely to yield exciting developments for some time to come.

The last topic treated in Chapter 7 concerns the role of collisions in quantum computation. Quantum information science is in its infancy, and it is too early to predict in which direction it will ultimately evolve. Nevertheless, the realization of quantum gate operations through the use of entangled atom states in optical lattices provides an instructive lesson on how to think about the nature of "information" and what must be the essential features of a quantum computer. Collisions provide the entangled states, and therefore their importance to future developments will remain crucial.

The starting point for this book was a *Review of Modern Physics* article (*Rev. Mod. Phys.*, 71: 1–85, 1999) co-authored by Paul Julienne, Vanderlei Bagnato, Sergio Zilio and me. That article reviewed developments in cold collisions through to the end of 1997, at the point where the Bose–Einstein condensate, first achieved in 1995, was just beginning to reveal its fascinating properties. Since then there has been an explosion of new physics in the control and manipulation of quantum-degenerate condensates (including literally exploding condensates!). The purpose of this book then is to not only update developments in the more established lines of cold collision research since the appearance of that article, but also to describe the important new role of collisions in the quantum-degenerate gases.

The author of this book owes a great debt of gratitude to his former co-authors of the *Review of Modern Physics* article mentioned above. Much of the theory presentation in Chapters 2 and 7 of this book is drawn from the contributions of Paul Julienne to that article, and the author has benefitted from many enlightening discussions with his colleagues at the Instituto de Física at the Universidade de São Paulo in São Carlos, Brazil and at the Université Paul Sabatier here in Toulouse. Whatever errors remain here are of course the sole responsibility of the present author.

# 1

# General introduction

In the 1980s the first successful experiments [312] and theory [98], demonstrating that light could be used to cool and confine atoms to submillikelvin temperatures, opened several exciting new chapters in atomic, molecular, and optical (AMO) physics. Atom interferometry [6, 8], matter–wave holography [294], optical lattices [192], and Bose–Einstein condensation in dilute gases [18, 95] all exemplified significant new physics where collisions between atoms cooled with light play a pivotal role. The nature of these collisions has become the subject of intensive study not only because of their importance to these new areas of AMO physics but also because their investigation has led to new insights into how cold collision spectroscopy can lead to precision measurements of atomic and molecular parameters and how radiation fields can manipulate the outcome of a collision itself. As a general orientation Fig. 1.1 shows how a typical atomic de Broglie wavelength varies with temperature and where various physical phenomena situate along the scale. With de Broglie wavelengths on the order of a few thousandths of a nanometer, conventional gas-phase chemistry can usually be interpreted as the interaction of classical nuclear point particles moving along potential surfaces defined by their associated electronic charge distribution. At one time liquid helium was thought to define a regime of cryogenic physics, but it is clear from Fig. 1.1 that optical and evaporative cooling have created "cryogenic" environments below liquid helium by many orders of magnitude. At the level of Doppler cooling and optical molasses[1] the de Broglie wavelength becomes comparable to or longer than the chemical bond, approaching the length of the cooling optical light wave. Here we can expect wave and relativistic effects such as resonances, interferences, and interaction retardation to become important. Following Suominen [370], we will term the Doppler cooling and optical molasses temperature range [407], roughly between 1 mK and 1 µK, the regime of cold collisions. Most collision phenomena at this level are studied in the presence of one or more light fields used to confine the atoms and to probe their interactions. Excited quasimolecular states often play an important role. Below about 1 µK, where evaporative cooling and

---

[1] A good introduction to the early physics of laser cooling and trapping can be found in two special issues of the *Journal of the Optical Society of America* B. These are: The mechanical effects of light, *J. Opt. Soc. Am.* B **2**, No. 11, November 1985 and Laser cooling and trapping of atoms, *J. Opt. Soc. Am.* B **6**, No. 11, November 1989. Two more recent reviews [7, 271] update a decade of developments since the early work recounted in the *J. Opt. Soc. Am.* B special issues.

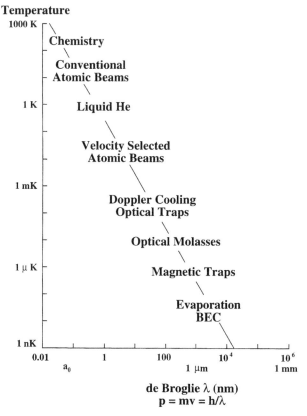

**Fig. 1.1.** Illustrative plot of various physical phenomena along a scale of temperature (energy divided by Boltzmann's constant $k_B$) plotted against the de Broglie wavelength for atomic sodium.

Bose–Einstein condensation[2] (BEC) become the focus of attention, the de Broglie wavelength grows to a scale comparable to the mean distance separating atoms at the critical condensation density; quantum degenerate states of the atomic ensemble begin to appear. In this regime ground-state collisions only take place through radial (not angular) motion and are characterized by a phase shift, or scattering length, of the ground-state wavefunction. Since the atomic translational energy now lies below the kinetic energy transferred to an atom by recoil from a scattered photon, light can play no role; and collisions occur in a temperature range from $1\ \mu K \rightarrow 0$ and in the dark. These collisions are termed ultracold. The terms "cold" and "ultracold" have been used in various ways in the past but now we adopt the terminology of Suominen [370] in distinguishing between the two. This book will recount progress in understanding collision processes in both the cold and ultracold domains.

---

[2] For an introduction to research in alkali-atom Bose–Einstein condensation see the special issue on Bose–Einstein condensation in the *Journal of Research of the National Institute of Standards and Technology* **101,** No. 4, July–August 1996.

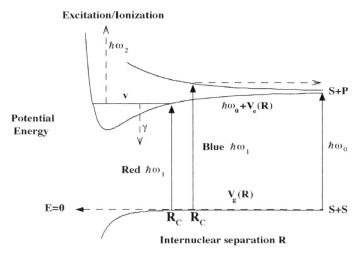

**Fig. 1.2.** Schematic of a cold collision. Light field $\hbar\omega_1$, red detuned with respect to the $S+P$ asymptote, excites the quasimolecule in a free–bound transition around the Condon point $R_C$ and can lead to excitation or ionization with the absorption of a second photon $\hbar\omega_2$. Light field $\hbar\omega_1$, blue detuned with respect to the $S+P$ asymptote, prevents atoms from approaching significantly beyond the Condon point. Blue-detuned excitation leads to optical shielding and suppression of inelastic and reactive collision rates.

We start with cold collisions because many experiments carried out in optical traps and optically slowed atomic beams take place in this temperature range. In 1986, soon after the first successful experiments reported optical cooling and trapping in alkali gasses, J. Vigué published a paper discussing the possible consequences of binary collisions in a cold or ultracold gaseous medium [415]. Figure 1.2 shows schematically the general features of a cold binary collision. Two S ground-state atoms interact at long range through electrostatic dispersion forces, and approach along an attractive $C_6/R^6$ potential. The first excited states, correlating to an $S+P$ asymptote and separated from the ground state by the atomic excitation energy $\hbar\omega_0$, interact by resonance dipole–dipole forces and approach either along an attractive or repulsive $C_3/R^3$ curve. If the colliding atoms on the ground-state potential encounter a *red-detuned* optical field $\hbar\omega_1$, the probability of a free–bound transition to some vibration–rotation level $v$ of the attractive excited state will maximize around the Condon point $R_C$, where $\hbar\omega_1$ matches the potential difference. The quasimolecule finds itself photoassociated and vibrating within the attractive well. A second field $\hbar\omega_2$ can then further excite or even ionize the photoassociated quasimolecule, or it can relax back to some distribution of the continuum and bound levels of the ground state with a spontaneous emission rate $\Gamma$. Photoassociation and subsequent inelastic processes comprise the discussion of Chapters 4 and 5. If the atoms colliding on the ground-state potential interact with a *blue-detuned* optical field $\hbar\omega$, the probability of transition to some continuum level of the repulsive excited state will maximize around the Condon point $R_C$, which will also be quite close to the turning point of the nuclear motion on the repulsive excited state. The atoms approach no further and begin to separate along the repulsive curve leading to the

S + P asymptote. The blue-detuned field "shields" the atoms from further interaction at more intimate internuclear separation and "suppresses" the rate of various inelastic and reactive processes. Optical shielding and suppression is the topic of Chapter 6. Cold collisions in the presence of optical fields or in the ground state reveal new physics in domains where atomic scales of length, time, and spectral line width are reversed from their conventional relations. In addition to the atomic de Broglie wave increasing to a length several hundred times that of the chemical bond, the collisional interaction time grows to several times the spontaneous emission lifetime, and the inhomogeneous Doppler line width at cold and ultracold temperatures narrows to less than the natural width of an atomic dipole transition. The narrow, near-threshold continuum state distribution means that at most a few partial waves will contribute to a scattering event. Averaging therefore does not obscure matter wave effects such as resonances, nodes and antinodes in scattering wavefunction amplitudes, and channel-opening threshold laws. In the cold collision regime, Doppler broadening is narrow compared to the radiative natural width; and therefore permits ultrahigh-precision, free–bound molecular spectroscopy and efficient participation of the entire atom ensemble in the excitation process. Long collision duration ensures interacting partners sufficient time to exchange multiple photons with modes of an externally applied radiation field. The frequency, intensity, and polarization of the optical field can in turn modify effective interaction potentials and control the probability of inelastic, reactive, and elastic final product channels.

Three questions have motivated significant developments in cold collisions:

(1)  How do collisions lead to loss of atom confinement in traps?
(2)  How can photoassociation spectroscopy yield precision measurements of atomic properties and insight into the quantum nature of the scattering process itself?
(3)  How can optical fields be used to control the outcome of a collisional encounter?

This third question has had an enduring appeal to chemical physicists, and we review here some of the early history to put current developments in perspective and to emphasize the importance of the ultracold regime. In the decade of the 1970s, after the development of the $CO_2$ laser, researchers in atomic and chemical physics immediately thought to use it to influence or control inelastic processes and chemical reaction. However, early attempts to induce reactivity by exciting well-defined, localized molecular sites such as double bonds or functional groups failed because the initial optical excitation diffused rapidly into the rotations, vibrations, and torsional bending motions of the molecular nuclei. The unfortunate result was that the infrared light of the $CO_2$ laser essentially heated the molecules much as the familiar and venerable Bunsen burner. Enthusiasm for laser-controlled chemistry cooled when researchers realized that this rapid energy diffusion throughout the molecular skeleton blocked significant advance along the road to optical control of reactivity. In the early 1980s the development of the pulsed dye laser, tunable in the visible region of the spectrum, together with new proposals for "radiative collisions," in which the electrons of the molecule interact with the light rather than the nuclei, revived interest. A second round of experiments achieved some success in optically transforming reactants to products; but in general the high peak powers necessary to enhance reactivity significantly interfered with the desired effects by inducing nonlinear, multiphoton excitation and ionization in the molecule itself. The necessity for high peak power in turn arose from two crucial factors: (1) Doppler broadening at ambient temperature permits only a few per cent of all molecular collisions to interact with an applied optical field mode so the product "yield" is low, and (2) the optical field had to influence the strong chemical binding interaction in order to affect atomic behavior during the collisional encounter. This requirement implied the need for power densities greater than a few megawatts per square centimeter. Power densities of this order are well above the threshold for multiphoton absorption and ionization, and these processes quickly convert atoms or molecules from a neutral gas into an ionized plasma. It appeared depressingly difficult to control collisions with light without first optically destroying them.

However, atomic deceleration and optical cooling brightened this discouraging picture, by narrowing the inhomogeneous Doppler broadening to less than a natural line width of an atomic transition and transferring the optical–particle interaction from the "chemical" zone of strong wavefunction overlap to an outer region where weak electrostatic terms characterize the collision. In this weakly interacting outer zone only hundreds of milliwatts per square centimeter of optical power density suffice to profoundly alter the inelastic and reactive collision rate constant. Furthermore, although a conventional atomic collision lasts only a few hundred femtoseconds, very short compared to the tens of nanoseconds required before excited molecules or atoms spontaneously emit light, in the ultracold regime particles move much more slowly, taking up to hundreds of nanoseconds to complete a collisional encounter. The long collision duration leaves plenty of time for the two interacting partners to absorb energy from an external radiation field or to emit energy by spontaneous or stimulated processes.

To the three earlier questions motivating studies of cold collisions may now be added a fourth relevant to the ultracold regime: what role do collisions play in the attainment and behavior of boson and fermion gases in the quantum degenerate regime? Quantum statistical effects have been observed and studied in the superfluidity of liquid helium and in the phenomena of metallic and high-temperature superconductivity. These dramatic and significant manifestations of quantum collective effects, are nevertheless difficult to study at the atomic level because the particles are condensed and strongly interacting. Observation and measurement in weakly interacting dilute gases, however, relate much more directly to the simplest, microscopic models of this behavior. The differences between model "ideal" quantum gases and "real" quantum gases begin with binary interactions between the particles, and therefore the study of ultracold collisions is a natural point of departure for investigation. Collisions determine two crucial aspects of BEC experiments: (1) the evaporative cooling rate necessary for the attainment of BEC depends on the elastic scattering cross section, proportional to the square of the s-wave scattering length, and (2) the sign of the scattering length indicates the stability of the condensate: positive scattering lengths lead to large stable condensates while negative scattering lengths do not. The ability to produce condensates by sympathetic cooling also depends critically on the elastic and inelastic (loss and heating) rate constants among different states of the colliding partners. Although a confined Bose atom condensate bears some analogy to an optical cavity (photons are bosons), the atoms interact through collisions; and these collisions limit the coherence length of any "atom laser" coupled out of the confining "cavity" or BEC trap. Another important point, relating back to optical control, is that the amplitude and sign of the scattering length depends sensitively on the fine details of the ground potentials. The possibility of manipulating these potentials and consequently collision rates with external means, using optical, magnetic, or radio-frequency fields, holds the promise of tailoring the properties of the quantum gas to enhance stability and coherence. Finally it now appears that the limiting loss process for dilute gaseous Bose–Einstein condensates are three-body collisions that also measure the third-order coherence, thus providing a critical signature of true quantum statistical behavior. Understanding the quantum statistical collective behavior of ultracold dilute gases will drive research in ultracold collisions for years to come.

# 2

# Introduction to cold collision theory

## 2.1  Basic concepts of scattering theory

Let us first consider some of the basic concepts that are needed to describe the collision of two ground-state atoms. We initially consider the collision of two distinguishable, structureless particles $a$ and $b$ with interaction potential $V_g(R)$ moving with relative momentum $\mathbf{k}$, where $\mathbf{R}$ is the vector connecting $a$ and $b$. We will generalize below to the cases of identical particles and particles with internal structure. The collision energy is

$$E = \frac{\hbar^2 k^2}{2\mu},$$

where $\mu$ is the reduced mass of the two particles. If there is no interaction between the particles, $V_g = 0$, the wavefunction describing the relative motion of the two particles in internal states $|0_a\rangle$ and $|0_b\rangle$ is

$$\Psi_g^+(\mathbf{R}) = e^{i\mathbf{k}\cdot\mathbf{R}}|0_a 0_b\rangle. \tag{2.1}$$

If the interaction potential is nonzero, the collision between the particles results in a scattered wave, and at large $R$ beyond the range of the potential the wavefunction is represented as

$$\Psi_g^+(\mathbf{R}) \sim \left[ e^{i\mathbf{k}\cdot\mathbf{R}} + \frac{e^{ikR}}{R} f(E, \hat{\mathbf{k}}, \hat{\mathbf{k}}_s) \right] |0_a 0_b\rangle, \tag{2.2}$$

where $\hat{\mathbf{k}}$ is a unit vector indicating the direction of $\mathbf{k}$ and $\hat{\mathbf{k}}_s$ is a unit vector indicating the direction of the scattered wave with amplitude $f(E, \hat{\mathbf{k}}, \hat{\mathbf{k}}_s)$. The overall effect of the collision is described by a cross section $\sigma(E)$. In a gas cell, for which all directions $\hat{\mathbf{k}}$ are possible, $\sigma(E)$ is determined by integrating over all scattered directions and averaging over all values of initial $\hat{\mathbf{k}}$:

$$\sigma(E) = \int_{4\pi} \frac{d\hat{\mathbf{k}}}{4\pi} \int_{4\pi} d\hat{\mathbf{k}}_s \, |f(E, \hat{\mathbf{k}}, \hat{\mathbf{k}}_s)|^2. \tag{2.3}$$

Since $f$ has units of length, the cross section has units of length$^2$ = area. Equation 2.3 simplifies for a spherically symmetric potential, for which $f$ depends only on the angle $\theta$

between $\hat{\mathbf{k}}$ and $\hat{\mathbf{k}}_s$:

$$\sigma(E) = 2\pi \int_0^\pi d\theta \, \sin(\theta) |f(E, \theta)|^2. \tag{2.4}$$

The object of scattering theory is to calculate the scattering amplitude and cross section, given the interaction potentials between the two atoms. The first step in reducing the problem to practical computation is to introduce the partial wave expansion of the plane wave:

$$e^{i\mathbf{k}\cdot\mathbf{R}} = 4\pi \sum_{\ell=0}^\infty \sum_{m=-\ell}^\ell i^\ell Y_{\ell m}^*(\hat{\mathbf{k}}) Y_{\ell m}(\hat{\mathbf{k}}_s) j_\ell(kR), \tag{2.5}$$

where $Y_{\ell m}$ is a spherical harmonic and the function $j_\ell$ has the following form as $R \to \infty$:

$$j_\ell(kR) \sim \frac{\sin\left(kR - \frac{\pi}{2}\ell\right)}{kR}. \tag{2.6}$$

The complete wavefunction at all $R$ is also expanded in a partial wave series:

$$\Psi_g^+(\mathbf{R}) = 4\pi \sum_{\ell=0}^\infty \sum_{m=-\ell}^\ell i^\ell Y_{\ell m}^*(\hat{\mathbf{k}}) Y_{\ell m}(\hat{\mathbf{k}}_s) \frac{F_\ell^+(E, R)}{R} |0_a 0_b\rangle, \tag{2.7}$$

where $F_\ell^+(E, R)$ is determined from the Schrödinger equation,

$$\frac{d^2 F_\ell^+(E, R)}{dR^2} + \frac{2\mu}{\hbar^2} \left[ E - V_g(R) + \frac{\hbar^2 \ell(\ell+1)}{2\mu R^2} \right] F_\ell^+(E, R) = 0. \tag{2.8}$$

By imposing the following boundary condition on $F_\ell^+(E, R)$ as $R \to \infty$, the asymptotic wavefunction has the desired form, Eq. 2.2, representing an incident plane wave plus a scattered wave:

$$\frac{F_\ell^+(E, R)}{R} \sim \sin\left(kR - \frac{\pi}{2}\ell + \eta_\ell\right) \frac{e^{i\eta_\ell}}{kR} \tag{2.9}$$

$$\sim j_\ell(kR) + \frac{i}{2} \frac{e^{i(kR - \frac{\pi}{2}\ell)}}{kR} T_\ell(E). \tag{2.10}$$

Here $\eta_\ell$ is the phase shift induced by the interaction potential $V_g(R)$, and $T_\ell(E) = 1 - e^{2i\eta_\ell}$ is the T-matrix element from which the amplitude of the scattered wave is determined,

$$f(E, \hat{\mathbf{k}}, \hat{\mathbf{k}}_s) = \frac{2\pi i}{k} \sum_{\ell=0}^\infty \sum_{m=-\ell}^\ell i^\ell Y_{\ell m}^*(\hat{\mathbf{k}}) Y_{\ell m}(\hat{\mathbf{k}}_s) T_\ell(E). \tag{2.11}$$

Using the simpler form of Eq. 2.4, the cross section becomes

$$\sigma(E) = \frac{\pi}{k^2} \sum_{\ell=0}^\infty (2\ell+1) |T_\ell(E)|^2 \tag{2.12}$$

$$= \frac{4\pi}{k^2} \sum_{\ell=0}^\infty (2\ell+1) \sin^2 \eta_\ell. \tag{2.13}$$

The cross section has a familiar semiclassical interpretation. If the interaction potential $V_g(R)$ vanishes, the particle trajectory is a straight line with relative angular momentum $\mathbf{R} \times \mathbf{p} = bp$, where $p$ is linear momentum and $b$ is the distance of closest approach. If we take the angular momentum to be

$$bp = \hbar\sqrt{\ell(\ell+1)} \approx \hbar\left(\ell + \frac{1}{2}\right),$$

where the "classical" $\ell$ here is not quantized, then $b$ is the classical turning point of the repulsive centrifugal potential $\hbar^2\ell(\ell+1)/2\mu R^2$ in Eq. 2.8. The semiclassical expression for the cross section, analogous to Eq. 2.12, has the form of an area,

$$\sigma(E, \text{semiclassical}) = 2\pi \int_0^\infty bP(b, E)\,db \qquad (2.14)$$

weighted by $P(b, E) = |T_\ell(E)|^2$. An important feature of cold collisions is that only a very few values of $\ell$ can contribute to the cross section, because the classical turning points of the repulsive centrifugal potential are at large values of $R$. Only collisions with the very lowest $\ell$ values allow the atoms to get close enough to one another to experience the interatomic interaction potential. We will discuss below the specific quantum properties associated with discrete values of $\ell$. Semiclassical theory is useful for certain types of trap loss collisions in relatively warm traps where light is absorbed by atom pairs at very large $R$, but a quantum treatment always becomes necessary at sufficiently low collision energy.

In the simple introduction above, the cross section in Eq. 2.13 represents elastic scattering, for which the internal states of the particles do not change, and their relative kinetic energy $E$ is the same before and after the collision. In general, the atoms will have nonzero internal angular momentum due to hyperfine structure in the case of alkali atoms or due to electronic structure in the case of rare gas metastable atoms. In a field-free region, these internal states are characterized by total angular momentum $F$ and projection $M$ on a space-fixed quantization axis. Often an external field, either a magnetic, optical, or radio-frequency electromagnetic field, is present in cold collision experiments; and these $FM$ states are modified by the external field. Species such as hydrogen and alkali atoms, which have $^2S$ ground states and nonvanishing nuclear spin $I$, have two hyperfine components with $F = I + \frac{1}{2}$ and $F = I - \frac{1}{2}$. Figure 2.1 shows the Zeeman splitting of the two $F = 1$ and $F = 2$ hyperfine components of ground-state $^{87}$Rb atoms, and Fig. 2.2 shows the ground-state interaction potentials for two interacting Rb atoms. If the atomic field-modified states, commonly called field-dressed states, are represented by $|\alpha_i\rangle$ for atom $i = a, b$, then the general collision process represents a transition from the state $|\alpha_a\rangle\,|\alpha_b\rangle$ to the state $|\alpha'_a\rangle|\alpha'_b\rangle$, represented by the general transition amplitude

$$f(E, \hat{\mathbf{k}}, \hat{\mathbf{k}}_s, \alpha_a\alpha_b \rightarrow \alpha'_a\alpha'_b)$$

and the T-matrix element

$$T(E, \alpha_a\alpha_b \rightarrow \alpha'_a\alpha'_b).$$

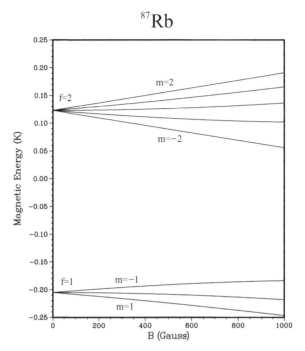

**Fig. 2.1.** Ground hyperfine levels of the $^{87}$Rb atom versus magnetic field strength. The Zeeman splitting manifolds are evident. The energy has been divided by the Boltzmann constant in order to express it in temperature units.

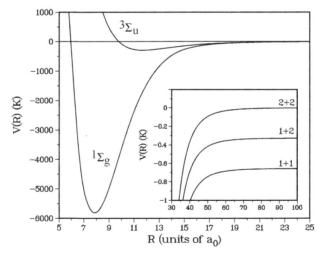

**Fig. 2.2.** The ground-state potential energy curves of the Rb$_2$ molecule. The potentials have been divided by the Boltzmann constant in order to express them in units of temperature. The full figure shows the short-range potentials on the scale of chemical bonding. The inset shows a blowup at long range, showing the separated atom hyperfine levels $F_a + F_b = 1 + 1, 1 + 2,$ and $2 + 2$. The upper two potentials in the inset correlate adiabatically with $^3\Sigma_u$.

These are the most complete transition amplitudes that describe an observable transition, specified in terms of quantities that can be selected experimentally. The most complete T-matrix elements that can be calculated from the multicomponent version of Eq. 2.8 are also specified by the initial and final partial waves:

$$T(E, \alpha_a \alpha_b \ell m \rightarrow \alpha'_a \alpha'_b \ell' m').$$

There are three distinct physical axes which define these amplitudes: a space-fixed axis that defines the space projection quantum numbers $M_a$ and $M_b$, the asymptotic direction of approach $\hat{\mathbf{k}}$, and the asymptotic direction of separation $\hat{\mathbf{k}}_s$. In beam experiments all three axes can be different. Almost all the work on cold atom collisions is carried out in a homogeneous gas, where neither $\hat{\mathbf{k}}$ nor $\hat{\mathbf{k}}_s$ are selected or measured, and a cell average cross section defined as in Eq. 2.12 for the general $\sigma(E, \alpha_a \alpha_b \rightarrow \alpha'_a \alpha'_b)$ is appropriate. Some experiments, especially in the context of Bose–Einstein condensation, have a well-defined local space-fixed axis, and select particular values of $M_a$ and $M_b$, but most of the work we will describe is for unpolarized gases involving an average over $M_a$ and $M_b$.

Instead of cross sections, it is usually preferable to give a rate coefficient for a collision process. The rate coefficient is directly related to the number of collision events occurring in a unit time in a unit volume. Consider the reaction

$$\alpha_a + \alpha_b \rightarrow \alpha'_a + \alpha'_b, \tag{2.15}$$

where the quantum numbers are all assumed to be different. If the density of species $\alpha_i$ in a cell is $n_{\alpha_i}$, the rate of change of the density of the various species due to collision events is

$$\frac{n_{\alpha_a}}{dt} + \frac{n_{\alpha_b}}{dt} = -\frac{n_{\alpha'_a}}{dt} - \frac{n_{\alpha'_b}}{dt} = K(T, \alpha_a \alpha_b \rightarrow \alpha'_a \alpha'_b) n_{\alpha_a} n_{\alpha_b}. \tag{2.16}$$

The rate coefficient $K(T, \alpha_a \alpha_b \rightarrow \alpha'_a \alpha'_b)$ is related to the cross section through

$$K(T, \alpha_a \alpha_b \rightarrow \alpha'_a \alpha'_b) = \langle \sigma(E, \alpha_a \alpha_b \rightarrow \alpha'_a \alpha'_b) v \rangle, \tag{2.17}$$

where the angled brackets imply an average over the distribution of relative collision velocities $v$.

The general theory of collisions of degenerate species is well understood. The basic multichannel theory can be found, for example, in Mies [274, 275] or Bayless et al. [38]. There is a considerable body of work on the theory of collisions of cold spin-polarized hydrogen atoms, as a consequence of the quest to achieve Bose–Einstein condensation in such a system. This is discussed in Chapter 7. The Eindhoven group led by B. Verhaar has been especially active in developing the theory. The paper by Stoof, Koelman, and Verhaar [365] gives an excellent introduction to the subject. The theory of cold collisions in external magnetic fields and optical or radio-frequency electromagnetic fields has also been developed, as we will describe in more detail in the following chapters of this book.

It is important to distinguish between two different kinds of collisions: elastic and inelastic. As mentioned above, an elastic collision is one in which the quantum states $\alpha_a$, $\alpha_b$ of each atom remain unchanged by the collision. These collisions exchange momentum, thereby aiding the thermalizing of the atomic sample. These are "good" collisions that do not destroy the trapped states, and they are necessary for the process of evaporative cooling that we will describe later. An inelastic collision is one in which one (or more) of these two quantum numbers changes in the collision. Most cold collision studies have dealt with inelastic events instead of elastic ones, that is, the collision results in hot atoms or untrapped species or even ionic species. As we will see in the next section, the quantum threshold properties of elastic and inelastic collisions are very different.

The basic difference between collisions of different atomic species and identical atomic species is the need to symmetrize the wavefunction with respect to exchange of identical particles in the latter case. Other than this symmetrization requirement, the theory is the same for the two cases. Symmetrization has two effects: the introduction of factors of 2 at various points in the theory, and the exclusion of certain states since they violate the exchange symmetry requirement. Such symmetry restrictions are well known in the context of diatomic molecular spectroscopy, leading, for example, to ortho and para species of molecular hydrogen and to every other line being missing in the absorption spectrum of molecular oxygen, due to the zero nuclear spin of the oxygen atom [171]. In the case of atomic collisions of identical species, if the two quantum numbers are also identical, $\alpha_a = \alpha_b$, only even partial waves $\ell$ are possible if the particles are composite bosons, and only odd partial waves are possible if the particles are composite fermions. If the two quantum numbers are not identical, $\alpha_a \neq \alpha_b$, both even and odd partial waves can contribute to collision rates. The effect of this symmetry is manifestly present in photoassociation spectra, where for example, half the number of lines appear in a doubly spin polarized gas (where all atoms are in the same quantum state) in contrast with an unpolarized gas (where there is a distribution of quantum states). These spectra are described in Chapter 5.

Stoof, Koelman, and Verhaar [365] give a good discussion of how to modify the theory to account for exchange symmetry of identical particles. Essentially, they set up the states describing the separated atoms, the so-called channel states of scattering theory, as fully symmetrized states with respect to particle exchange. T-matrix elements and event rate coefficients, defined as in Eq. 2.17, are calculated conventionally for transitions between such symmetrized states. The event rate coefficients are given by

$$K(\{\gamma\delta\} \to \{\alpha\beta\}) = \left\langle \frac{\pi\hbar}{\mu k} \sum_{\ell'm'} \sum_{\ell m} |T(E, \{\gamma\delta\}\ell'm', \{\alpha\beta\}\ell m)|^2 \right\rangle, \tag{2.18}$$

where the braces $\{\cdots\}$ signify symmetrized states, and the T-matrix as defined in this review is related to the unitary S-matrix by $\mathbf{T} = \mathbf{1} - \mathbf{S}$. Then collision rates are unambiguously given by

$$\frac{dn_\alpha}{dt} = \sum_\beta \sum_{\{\gamma\delta\}} (1 + \delta_{\alpha\beta})[K(\gamma\delta \to \alpha\beta)n_\gamma n_\delta - K(\alpha\beta \to \delta\gamma)n_\alpha n_\beta]. \tag{2.19}$$

This also works for the case of elastic scattering of identical particles in identical quantum states: $K(\alpha\alpha \to \alpha\alpha)$ must be multiplied by a factor of 2 to get the rate of momentum transfer ($\hat{\mathbf{k}}$ scatters to $\hat{\mathbf{k}}_s \neq \hat{\mathbf{k}}$) since two atoms scatter per collision event. Gao [145] has also described the formal theory for collisions of cold atoms taking into account identical particle symmetry.

## 2.2  Quantum properties as energy approaches zero

Cold and ultracold collisions have special properties that make them quite different from conventional room temperature collisions. This is because of the different scales of time and distance involved. The effect of the long collision time is discussed in Chapter 4. Here we examine the consequence of the long de Broglie wavelength of the colliding atoms. The basic modification to collision cross sections when the de Broglie wavelength becomes longer than the range of the potential was described by Bethe [43] in the context of cold neutron scattering, and has been widely discussed in the nuclear physics literature [107, 444]. Such quantum threshold effects only manifest themselves in neutral atom collisions at very low temperature, typically $\ll 1$ K [202, 204], but they play a very important role in the regime of laser cooling and evaporative cooling that has been achieved.

Figure 1.1. gives a qualitative indication of the range of $\lambda$ as temperature is changed. Even at the high-temperature end of the range of laser cooling, $\lambda$ on the order of 100 $a_0$ is possible, and at the lower end of evaporative cooling, $\lambda > 1$ μm (20 000 $a_0$). These distances are much larger than the typical lengths associated with chemical bonds, and the delocalization of the collision wavefunction leads to characteristic behavior where collision properties scale as some power of the collision momentum $k = \sqrt{2\mu E} = 2\pi/\lambda$ as $k \to 0$, depending on the inverse power $n$ of the long-range potential, which varies as $R^{-n}$.

In the case of elastic scattering, Mott and Massey [295] show that the phase shift $\eta_\ell$ in Eqs. 2.9 and 2.13 has the following property as $k \to 0$: if $2\ell < n - 3$,

$$\lim_{k \to 0} k^{2\ell+1} \cot \eta_\ell = -\frac{1}{A_\ell}, \tag{2.20}$$

where $A_\ell$ is a constant, whereas if $2\ell > n - 3$,

$$\lim_{k \to 0} k^{n-2} \cot \eta_\ell = \text{constant.} \tag{2.21}$$

For neutral ground-state atoms, this ensures that the phase shift vanishes at least as fast as $k^3$ for all $\ell \geq 1$. Thus all contributions to the cross section vanish when $k$ becomes sufficiently small, except the contribution from the s-wave, $\ell = 0$. Since the s-wave phase shift varies as $-kA_0$ as $k \to 0$, we see from Eq. 2.13 that the elastic scattering cross section for identical particles approaches

$$\sigma(E) \to 8\pi A_0^2, \tag{2.22}$$

where the factor of 8 instead of 4 occurs due to identical particle symmetry. Thus, the cross section for elastic scattering becomes constant in the low-energy limit. The quantity $A_0$ is the s-wave scattering length, an important parameter in the context of Bose–Einstein condensation. Note that the rate coefficient for elastic scattering vanishes as $T^{1/2}$ in the limit of low temperature, since $K = \langle \sigma v \rangle$.

Figure 2.3 illustrates the properties of a very low collision energy wavefunction, taken for a model potential with a van der Waals coefficient $C_6$ and mass characteristic of Na

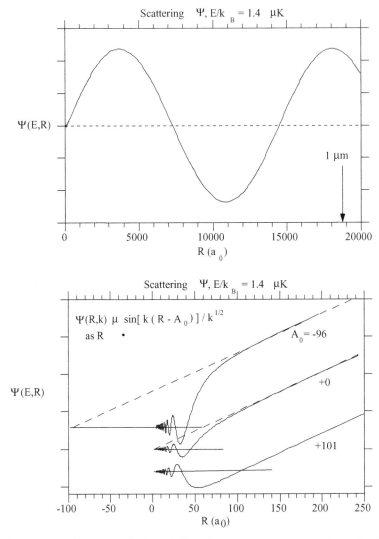

**Fig. 2.3.** The upper panel illustrates the long de Broglie wave at long range, on the scale of 1 μm. The lower panel shows a blowup of the short-range wavefunction for the case of three different potentials, with three different scattering lengths, negative, zero, and positive.

atom collisions. The upper panel illustrates the wavefunction at long range, on the scale of 1 μm. The lower panel shows a blowup of the short-range wavefunction for the case of three different potentials, with three different scattering lengths, negative, zero, and positive. The figure shows the physical interpretation of the scattering length, as the effective point of origin of the long-wavelength asymptotic sine wave at $R = A_0$, since the long-range wavefunction is proportional to $\sin[k(R - A_0)]$.

Inelastic collisions have very different threshold properties than elastic ones. As long as the internal energy of the separated atoms in the exit channel is lower than for the entrance channel, so that energy is released in the collision, the transition matrix element varies as, [43], [202], [444]

$$T(E) \propto k^{\ell + \frac{1}{2}},$$  (2.23)

where $\ell$ is the entrance channel partial wave index. Using this form in Eq. 2.12 shows that the cross section vanishes at least as fast as $k$ for all $\ell > 0$, but it varies as $1/k$ for the s-wave. This variation (sometimes called the $1/v$ law) was given by [43] and is well known in nuclear physics. Although the cross section for an inelastic, energy-releasing, collision becomes arbitrarily large as $k \to 0$, the rate coefficient $K$ remains finite, and approaches a nonvanishing constant.

The range of $k$ where these limiting threshold laws become valid depends strongly on the particular species, and even on the specific quantum numbers of the separated atoms. Knowledge of the long-range potential alone does not provide a sufficient condition to determine the range of $k$ in which they apply. This range, as well as the scattering length itself, depends on the actual phase shift induced by the whole potential, and is very sensitive to uncertainties in the short-range part of the potential in the chemical bonding region [159]. On the other hand, a necessary condition for threshold law behavior can be given based solely on the long-range potential [202], [204]. This condition is based on determining where a semiclassical, Wentzel–Kramers–Brillouin (WKB) connection breaks down between the long-range asymptotic s-wave and the short-range wavefunction, which experiences the acceleration of the potential. Consider the ground-state potential $V_g(R)$ as a function of $R$. The long-range potential is

$$V_g(R) = -\frac{C_n}{R^n},$$  (2.24)

where $n$ is assumed to be $\geq 3$. Let us first define

$$E_Q = \frac{\hbar^2}{\mu} \left[ \left( 2\frac{n+1}{3} \right)^{2n} \left( \frac{n-2}{6n} \right)^n \left( \frac{2n+2}{n-2} \frac{\hbar^2}{\mu C_n} \right)^2 \right]^{\frac{1}{n-2}}$$  (2.25)

$$R_Q = \left( \frac{n-2}{2n+2} \frac{C_n}{E_Q} \right)^{\frac{1}{n}}.$$  (2.26)

The properties of the wavefunction $\Psi(E, R)$ depend on the values of $E$ and $R$ relative to $E_Q$ and $R_Q$. When $E \gg E_Q$, the energy is high enough that it is always possible to make a semiclassical WKB connection between the wavefunction in the long-range zone, $R \gg R_Q$, and the wavefunction in the short-range zone, $R \ll R_Q$. In this case, the WKB representation of the s-wave wavefunction is a good approximation at all $R$, and there are no special threshold effects. On the other hand, $E \ll E_Q$ satisfies the condition that there is some range of distance near $R \approx R_Q$ were the WKB approximation fails, so that no semiclassical connection is possible between long- and short-range wavefunctions. This is the regime where a quantum connection, with its characteristic threshold laws, is necessary. When $E \ll E_Q$ the wavefunction for $R \gg R_Q$ is basically the asymptotic wave with the asymptotic phase shift [199], whereas the wavefunction for $R \ll R_Q$ is attenuated in amplitude, relative to the normal WKB amplitude, by a factor proportional to $k^{\ell+1/2}$ [202].

**Table 2.1.** Characteristic threshold parameters $R_Q$ and $E_Q$.

| Species | $R_Q$ (Bohr) | $E_Q$ (mK) | Species | $R_Q$ (Bohr) | $E_Q$ (mK) |
|---------|------------|-----------|---------|------------|-----------|
| Li | 32 | 120 | He* | 34 | 180 |
| Na | 44 | 19 | Ne* | 40 | 26 |
| K | 64 | 5.3 | Ar* | 60 | 5.7 |
| Rb | 82 | 1.5 | Kr* | 79 | 1.6 |
| Cs | 101 | 0.6 | Xe* | 96 | 0.6 |

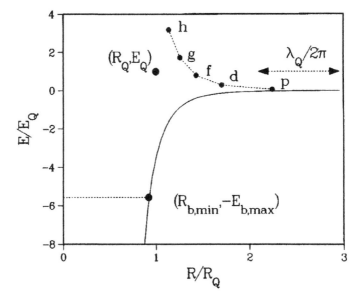

**Fig. 2.4.** The position of the maximum in the barrier due to the centrifugal potential for $\ell \neq 0$, as well as the maximum binding energy and minimum outer classical turning point of the last bound state in the van der Waals potential, all have identical scalings with $C_6$ and mass as $R_Q$ and $E_Q$, scaling as $(\mu C_6)^{1/4}$ and $\mu^{3/2} C_6^{1/2}$, respectively.

Julienne, Smith, and Burnett, [204], calculated the values of $E_Q$ and $R_Q$ for a number of alkali and rare gas metastable species that can be laser cooled (these are shown in Table 2.1.) The alkali values were based on known values of $C_6$ for the $n = 6$ van der Waals potentials, and estimates were made for the metastable rare gas $C_6$ coefficients, for which the weak $C_5/R^5$ quadrupole–quadrupole contribution to the potential for metastable rare gases was ignored in comparison to the $C_6$ contribution. The value of $R_Q$ is small enough that relativistic retardation corrections to the ground-state potential are insignificant. The $R_Q, E_Q$ parameters for each rare gas are similar to those of the neighboring alkali species in the periodic table. Figure 2.4 shows a plot of the ground-state potential expressed in units of $R_Q$ and $E_Q$. The position of the maximum in the barrier due to the centrifugal potential for $\ell \neq 0$, as well as the maximum binding energy and minimum outer classical turning point of the last bound state in the van der Waals potential, scale with $C_6$ and mass as $R_Q$ and $E_Q$ do, scaling as $(\mu C_6)^{1/4}$ and $\mu^{-3/2} C_6^{-1/2}$, respectively. These other scalable parameters are also indicated on Fig. 2.4, which gives a universal plot for any of the species in Table 2.1. Figure 2.4 also shows $\lambda/2\pi = 0.0857 R_Q$ for $E = E_Q$. The centrifugal barriers for the lowest few partial waves are lower than $E_Q$, but lie outside $R_Q$. The positions $R_C(\ell)$ and heights $E_C(\ell)$ of the centrifugal barriers are $R_C(\ell) = 2.67 R_Q/[\ell(\ell+1)]^{1/4}$ and $E_C(\ell) = 0.0193 E_Q [\ell(\ell+1)]^{3/2}$, respectively. The p- and d-wave barriers have heights of $0.055 E_Q$ and $0.28 E_Q$, respectively. $E_Q$ is small enough that several partial waves typically contribute to ground state collisions of the heavier species at typical magneto-optical trap (MOT) temperatures of 100 μK to 1 mK.

The distance $R_Q$, closely related to the mean scattering length $\bar{a} = 0.967 R_Q$ defined by Gribakin and Flambaum, [159], is solely a property of the long-range potential and atomic mass. On the other hand, the approach to the quantum threshold laws is in the range $0 < E \ll E_Q$, where the actual range depends on the effect of the whole potential, as measured by the threshold phase shift proportional to the actual s-wave scattering length $A_0$. Let $A_b$ be defined by $E_b = \hbar^2/2\mu A_b^2$, where $E_b$ is the position of the last bound state of the potential just below threshold ($A_b > 0$) or the position of a virtual bound state just above threshold ($A_b < 0$). When $E_b < E_Q$, the range over which the threshold laws apply can be estimated from $0 < k \le |A_b|^{-1}$, or $0 < E \ll E_b$. Thus, the threshold laws apply only over a range comparable to the lesser of $E_Q$ and $E_b$. In the limit where the last bound state is very near threshold, [159] finds that $A_0 = \bar{a} + A_b$; this reduces to the result of Mies et al. [277], $A_0 \approx A_b$, when $|A_0| \gg \bar{a}$. The special threshold properties of the last bound state in the potential has recently been described by Boisseau et al. [52] and Trost et al. [400]. The usual quantization rules in a long-range potential [242] do not adequately characterize the threshold limit of the last bound state. As we shall see in Section 2.2.2, the scattering length can be arbitrarily large, corresponding to the case where the last bound state is arbitrarily close to threshold, $E_b$ can be arbitrarily small. Therefore, the actual range over which the threshold law expressions in Eqs. 2.22 and 2.23 apply can in principle be less than the actual temperature range in cold or even ultracold experiments. In fact, the Cs atom in its doubly spin polarized hyperfine level seems not yet to be in the threshold limit even at a few μK [19], [239].

### 2.2.1   Relations between phase shift, scattering length, and bound states

In the ultracold regime, where only s-wave elastic scattering occurs, the magnitude and sign of the scattering length determines all the collisional properties. We will see in Chapter 7 the crucial importance of these properties to the stability and internal dynamics of Bose–Einstein condensates. It is worthwhile therefore to review the relationship between the phase shift, the scattering length, and how they behave in the presence of bound states just at the potential threshold. We follow the discussion by Joachain [193], adapting it to the notation already established.

For s-wave scattering we need only consider the radial part of the Schrödinger equation,

$$-\frac{\hbar^2}{2m}\left[\frac{1}{R^2}\frac{d}{dr}\left(R^2\frac{d}{dR}\right) - \frac{l(l+1)}{R^2}\right]\mathcal{R}_l(k, R) + V(R)\mathcal{R}_l(k, r) = E\mathcal{R}_l(k, R), \quad (2.27)$$

where $\mathcal{R}_l(k, R)$ are the radial solutions. The Schrödinger equation can be cast into a more tractable form by making two judicious substitutions. In place of $V(R)$ we use a "reduced" potential

$$U(R) = \frac{2m}{\hbar^2}V(R) \quad (2.28)$$

and introduce a new radial function

$$u_l(k, R) = R\,\mathcal{R}_l(k, R). \quad (2.29)$$

With these substitutions the Schrödinger radial equation takes on the simpler form

$$\left[\frac{d^2}{dR^2} + k^2 - \frac{l(l+1)}{R^2} - U(R)\right]u_l(k, R) = 0, \quad (2.30)$$

where we have used $E = \hbar^2 k^2/2\mu$. Now Eq. 2.30, with $U(R) = 0$, assumes the form of the spherical Bessel differential equation that has two sets of linearly independent solutions useful for scattering problems. The first set comprises the spherical Bessel and Neumann functions, $j_l(kR)$ and $n_l(kR)$, respectively. The second set is the spherical Hankel functions, $h_l^{(1)}$ and $h_l^{(2)}$. The Bessel and Neumann functions are real. Their asymptotic forms are

$$j_l(kR) \xrightarrow[kR \to \infty]{} \frac{1}{kR}\sin\left(kR - \frac{1}{2}l\pi\right) \quad (2.31)$$

$$n_l(kR) \xrightarrow[kR \to \infty]{} -\frac{1}{kR}\cos\left(kR - \frac{1}{2}l\pi\right). \quad (2.32)$$

The Hankel functions are complex exponential functions. Their asymptotic forms are

$$h_l^{(1)}(kR) \xrightarrow[kR \to \infty]{} -i\frac{e^{i(kR-1/2l\pi)}}{kR} \quad (2.33)$$

$$h_l^{(2)}(kR) \xrightarrow[kR \to \infty]{} i\frac{e^{-i(kR-1/2l\pi)}}{kR}. \quad (2.34)$$

The real and complex set of solutions are related by linear combinations reminiscent of the relations between ordinary sin and cos functions and their exponential representations.

Near the origin the Bessel function is a *regular* solution; its limiting value is unity for $l = 0$ and it vanishes for all higher partial waves. The Neumann function, as well as the two Hankel functions, are singular with a pole of order $l + 1$. As already indicated by Eq. 2.6, the $U(R) = 0$ solution to the radial equation must be $j_l(R)$ since it does not have unphysical singularities at the scattering center. When a scattering potential is added to Eq. 2.30, the general solution must be some linear combination of one of the two sets of linearly independent functions, the real set, $j_l$, $n_l$, or the complex set $h_l^{(1)}$, $h_l^{(2)}$. Here we choose the real set and assume that the potential has a *finite range* such that there is a distance from the scattering center $R = a$ where $U(R)$ becomes negligible. This condition applies to atom–atom or ion–atom collisions, but is not satisfied for collisions between charged species. We can write the general solution $u_l(kR)$ in this asymptotic region as a linear combination of the two asymptotic forms of $j_l(kR)$ and $n_l(kR)$,

$$u_l(k, R) = kR\left[C_l^{(1)}(k)j_l(kR) + C_l^{(2)}(k)n_l(kR)\right]. \tag{2.35}$$

Now nothing prevents us from writing the "mixing coefficients" as

$$C_l^{(1)}(k) = \cos \eta_l(k) \quad \text{and} \quad C_l^{(2)}(k) = -\sin \eta_l(k) \tag{2.36}$$

so that

$$\tan \eta_l(k) = -\frac{C_l^{(2)}}{C_l^{(1)}} \tag{2.37}$$

and

$$u_l(k, R) \xrightarrow[R \to \infty]{} C_l(k) \sin\left[kR - \frac{1}{2}l\pi + \eta_l(k)\right]. \tag{2.38}$$

Thus the effect of the scattering potential is to "mix in" some Neumann function to the general solution, and the result of this mixing is simply the phase shift $\eta_l(k)$. The factor $C_l(k)$ is a normalization constant for $u_l(k, R)$, and it is clear from the linear independence of $j_l(kR)$ and $n_l(kR)$ that

$$[C_l(k)]^2 = \left[C_l^{(1)}\right]^2 + \left[C_l^{(2)}\right]^2.$$

It is easy to show that the radial solution can also be written as

$$u_l(k, R) \xrightarrow[R \to \infty]{} \tilde{C}_l\left[\sin\left(kR - \frac{1}{2}\pi\right) - \tan \eta_l(k) \cos\left(kR - \frac{1}{2}\pi\right)\right] \tag{2.39}$$

and from Eq. 2.29 the solutions to the Schrödinger equation

$$R_l(k, R) \xrightarrow[R \to \infty]{} \tilde{C}_l(k)\left[j_l(kR) - \tan \eta_l(k)n_l(kR)\right]. \tag{2.40}$$

Now we make use of the *finite range* of the potential $U(R)$. At internuclear distances $R > a$ we will consider the potential to be zero so that the asymptotic form Eq. 2.40 is the relevant solution to the Schrödinger equation. At shorter distances we assume a well-defined potential form with known solutions $R_l$ for Eq. 2.30. At the boundary $R = a$ we apply joining conditions to the interior and exterior solutions. Smooth joining requires that the solutions and their derivatives match at the boundary $R = a$. A convenient way to express this matching is the *logarithmic derivative*:

$$\gamma_l = \left[\frac{dR_l/dR}{R_l}\right]_{R=a}. \tag{2.41}$$

Using Eq. 2.40 for the exterior solution we express the logarithmic derivative at the boundary as

$$\gamma_l = \frac{k\left[j'_l(ka) - \tan \eta_l(k)n'_l(ka)\right]}{j_l(ka) - \tan \eta_l(k)n_l(ka)}, \tag{2.42}$$

where $j'_l$, $n'_l$ indicate differentiation with respect to $R$. From this joining condition we obtain an expression for the phase shift in terms of the asymptotic solutions and $\gamma_l$,

$$\tan \eta_l(k) = \frac{kj'_l(ka) - \gamma_l(k)j_l(ka)}{kn'_l(ka) - \gamma_l(k)n_l(ka)}. \tag{2.43}$$

It can be shown [193] that in the limit $k \to 0$ the tangent of the phase shift takes on the limiting form (for $l = 0$)

$$\tan \eta_0 \underset{k \to 0}{\longrightarrow} -ka\left(\frac{a\hat{\gamma}_0}{1 + a\hat{\gamma}_0}\right), \tag{2.44}$$

where

$$\lim_{k \to 0} \gamma_l \to \hat{\gamma}_0.$$

We identify the factor

$$a\left(\frac{a\hat{\gamma}_0}{1 + a\hat{\gamma}_0}\right)$$

with the *scattering length* $A_0$ and write

$$A_0 = -\lim_{k \to 0} \frac{\tan \eta_0(k)}{k}. \tag{2.45}$$

Except for isolated points of *zero-energy resonances* (discussed in Section 2.2.2) the phase shift goes to zero with $k$. Therefore we can write

$$A_0 = -\lim_{k \to 0} \frac{\sin \eta_0(k)}{k}. \tag{2.46}$$

So from Eq. 2.13 the total scattering cross section at very low energy approaches

$$\sigma = 4\pi A_0^2. \tag{2.47}$$

In the case of identical particles, as already pointed out in Eq. 2.22, an extra factor of 2 must be inserted in the cross section formula.

### 2.2.2   Scattering length in a square-well potential

To illustrate the behavior of the phase shift and scattering length as the collision energy approaches zero, we calculate them for the simple case of an attractive square-well potential. We take

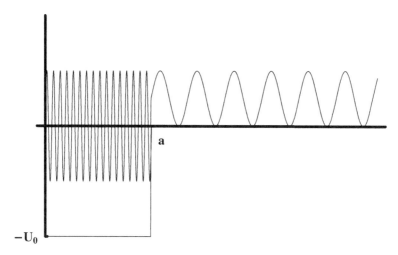

**Fig. 2.5.** Scattering from a 1-D "square-well" potential: the waves in the inner and outer regions must join smoothly at $R = a$. In reduced units the kinetic energy in the outer region is $k^2$ and in the inner region $\kappa^2$. The two are related by $\kappa = \sqrt{k^2 + U_0}$.

$$U(R) = -U_0 \qquad R < a$$
$$U(R) = 0 \qquad R > a.$$

In the interior region, for s-wave scattering, the radial equation will have the form

$$\left[ \frac{d^2}{dR^2} + \kappa^2 - U_0(R) \right] u_0(k, R) = 0, \tag{2.48}$$

where $\kappa = \sqrt{k^2 + U_0}$. The solution of the Schrödinger equation must look like the "regular" solution,

$$\mathcal{R}_0 = C_0 \, j_0(\kappa R).$$

In the exterior region we should have, according to Eq. 2.40,

$$\mathcal{R}_0(R) = \tilde{C}_0(k) \left[ j_0(kR) - \tan \eta_0 \, n_0(kR) \right].$$

This one-dimensional (1-D) scattering problem is depicted in Fig. 2.5. Application of the joining conditions at the boundary, Eq. 2.43, results in a expression for the phase shift

$$\eta_0 = -ka + \tan^{-1} \left( \frac{k}{\kappa} \tan \kappa a \right). \tag{2.49}$$

As $k \to 0$

$$\eta_0 \simeq ka \left( \frac{\tan \kappa a}{\kappa a} - 1 \right). \tag{2.50}$$

Figure 2.6 shows the behavior of the phase shift as the potential well depth increases. These periodic divergences in the phase shift are directly related to the appearance of bound states

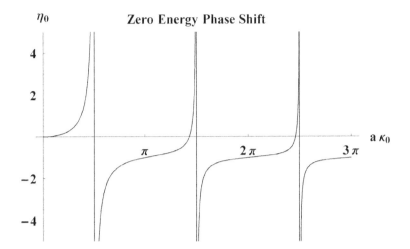

**Fig. 2.6.** Behavior of the phase shift as $a\kappa_0$ increases. The term $\kappa_0 = \sqrt{U_0}$ and is the limiting expression for $\kappa$ as the collision energy goes to zero. The phase shift diverges at $\pi/2$ (modulo $\pi$) as $a\kappa_0$ increases. For fixed potential width the increase of $a\kappa_0$ can be interpreted as increasing well depth.

in the potential. The condition for the appearance of bound states is that the matching conditions, Eq. 2.41, should be satisfied for a wave solution in the outer zone ($R > a$) that falls off exponentially with distance beyond $a$ instead of propagating indefinitely. Straightforward application of the logarithmic derivative matching shows that the condition for bound states is given by

$$\xi \cot \xi = -\zeta \tag{2.51}$$

with $\xi = \kappa a$ and $\zeta = ka$. Equation 2.51 is transcendental and does not have closed-form solutions. The position of the bound states can be found graphically, however, by plotting each side of the equation with respect to $\kappa a$ and $ka$ and determining where the functions intersect. Figure 2.7 illustrates this procedure. The key point is that in the limit $k \to 0$ the s-wave bound states appear at the same positions of $\kappa a$ as the divergences in the phase shift plotted in Fig. 2.6 In other words, phase shift divergence is a signature of the appearance of bound states in the attractive scattering potential. This condition is sometimes called the *zero-energy resonance*. Since the scattering length is simply related to the phase shift, it is not surprising that the scattering length also exhibits divergence when bound states appear. Figure 2.8 shows the behavior of the scattering length as a function of $\kappa a$ as the collision energy approaches zero. For fixed $a$ this variation shows how the scattering length varies with increasing potential well depth. When $U_0$ is too shallow to support any bound states, the scattering length is negative. At the appearance of the first bound state at threshold, $A_0$ diverges, changes sign and reappears as a positive scattering length. Increasing the potential depth still further continues to reduce $A_0$ until it passes smoothly through zero and again diverges negatively at the threshold for the second bound state.

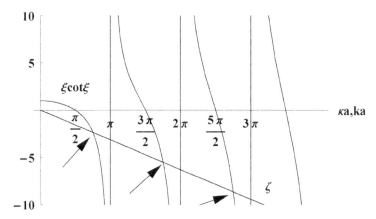

**Fig. 2.7.** Plot of the left-hand side of Eq. 2.51 versus $\kappa a$ and the right-hand side of Eq. 2.51 versus $\kappa a$. The points of intersection determine the positions of the bound states. Note that as $k \to 0$, the straight line $\zeta$ approaches the axis, and the bound states appear at $\kappa_0 a = \pi/2$ (modulo $\pi$).

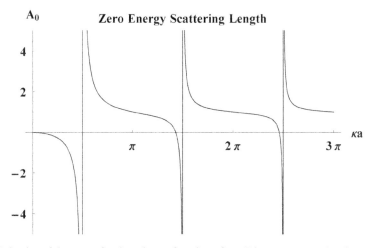

**Fig. 2.8.** Behavior of the scattering length as a function of $\kappa a$. Divergences are the signature for the appearance of bound states in the square well potential. The values of $\kappa a$ at which the divergences appear correspond to a collision energy approaching zero.

## 2.3 Collisions in a light field

Much of the work on collisions of cooled and trapped atoms has involved collisions in a light field. This is the subject of Chapters 4–6, of this book and been the subject of earlier reviews [419], [440]. Therefore, it is useful to provide an overview of some of the concepts that are used in understanding cold collisions in a light field. Figure 1.2. gives a schematic description of the key features of these collisions. Figure 1.2. shows ground and excited-state potentials, $V_g(R)$ and $V_e(R)$, and indicates the optical coupling between ground and excited states induced by a light field at frequency $\omega_1$, which is detuned by $\Delta = \omega_1 - \omega_0$ from the frequency of atomic resonance $\omega_0$. Detuning can be to the red or blue of resonance, corresponding to negative or positive $\Delta$, respectively. The figure illustrates that the excited-state potential, having a resonant-dipole form that varies as $C_3/R^3$, which can be attractive or repulsive, is of much longer range than the ground-state van der Waals potential. Figure 1.2. also indicates the Condon point $R_C$ for the optical transition. This is the point at which the quasimolecule comprised of the two atoms is in optical resonance with the light field, i.e.

$$V_e(R_C) - V_g(R_C) = \hbar\omega_1. \tag{2.52}$$

The Condon point is significant because in a semiclassical picture of the collision, it is the point at which the transition from the ground state to the excited state is considered to occur. The laser frequency $\omega_1$ can be readily varied in the laboratory, over a wide range of red and blue detuning. Thus, the experimentalist has at his disposal a way of selecting the Condon point and the upper state which is excited by the light. By varying the detuning from very close to atomic resonance to very far from resonance, the Condon point can be selected to range from very large $R$ to very small $R$.

Although the semiclassical picture implied by using the concept of a Condon point is very useful, a proper theory of cold collisions should be quantum mechanical, accounting for the delocalization of the wavefunction. If the light is not too intense, the probability of a transition, $P_{ge}(E, \omega_1)$, is proportional to a Franck–Condon overlap matrix element between ground- and excited-state wavefunctions:

$$P_{ge}(E, \omega_1) \propto |\langle\Psi_e(R)|\Omega_{eg}(R)|\Psi_g(E, R)\rangle|^2 \tag{2.53}$$

$$\approx |\Omega_{eg}(R_C)|^2 |\langle\Psi_e(R)|\Psi_g(E, R)\rangle|^2, \tag{2.54}$$

where the optical Rabi matrix element $\Omega_{eg}(R)$ measures the strength of the optical coupling. Julienne [199] showed that a remarkably simple reflection approximation to the Franck–Condon factor, based on expanding the integrand of the Franck–Condon factor about the Condon point, applies over a wide range of ultracold collision energies and red or blue detunings. This approximation, closely related to the usual stationary phase approximation of line broadening theory, shows that the Franck–Condon factor (in the cases where there is only one Condon point for a given state) is proportional to the ground-state wavefunction at the Condon point:

$$F_{ge}(E, \Delta) = |\langle\Psi_e(R)|\Psi_g(E, R)\rangle|^2 \tag{2.55}$$

$$= \frac{1}{D_C}|\Psi_g(E, R_C)|^2 \qquad \text{blue detuning} \tag{2.56}$$

$$= h v_v \frac{1}{D_C}|\Psi_g(E, R_C)|^2 \quad \text{red detuning.} \tag{2.57}$$

Here $D_C = |d(V_e - V_g)/dR|_{R_C}$ is the slope of the difference potential, and the wavefunctions have been assumed to be energy-normalized, that is, the wavefunctions in Eqs. 2.7–2.9 are multiplied by $\sqrt{2\mu k/\pi\hbar^2}$ so that

$$|\langle\Psi_g(E, R)|\Psi_g(E', R)\rangle|^2 = \delta(E - E'). \tag{2.58}$$

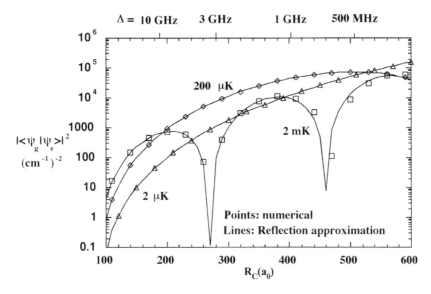

**Fig. 2.9.** The plots show the validity of the reflection approximation over a range of Condon points and collision energies for a typical case.

The formulas for red and blue detuning differ because they represent free–bound and free–free transitions, respectively; here $\nu_v$ is the vibrational frequency of the bound vibrational level excited in the case of red detuning. Figure 2.9 shows the validity of this approximation over a range of Condon points and collision energies for a typical case. We thus see that it is legitimate to use the semiclassical concept of excitation at a Condon point in discussing cold collisions in a light field.

The excited-state potential can be written at long range as [270],

$$V_e(R) = f\hbar\Gamma_A \left(\frac{\lambda_A}{2\pi R}\right)^3, \tag{2.59}$$

where $\lambda_A$ and $\Gamma_A$ are the respective transition wavelength and spontaneous decay width of the atomic resonance transition, and where $f$ is a dimensionless factor on the order of unity. Thus, in a MOT, where $|\Delta|$ is on the order of $\Gamma_A$, the Condon point will have a magnitude on the order of $\lambda_A/2\pi$, that is, on the order of $1000\,a_0$ for typical laser cooling transitions. This is the regime of the trap loss experiments described in Chapter 4. Optical shielding experiments are typically done with detuning up to several hundred $\Gamma_A$, corresponding to much smaller Condon points, on the order of a few hundred $a_0$. Photoassociation spectroscopy experiments use quite large detunings, and typically sample Condon points over a range from short-range chemical bonding out to a few hundred $a_0$. Because of the very different ranges of $\Delta$ and $R_C$ in the various kinds of cold collisions in a light field, a number of different kinds of theories and ways of thinking about the phenomena have been developed to treat the various cold collision processes. Since $R_C$ for MOT trap loss collisions is typically much larger than the centrifugal barriers illustrated in Fig. 2.4 many partial waves contribute to the collision, the quantum threshold properties are not evident at all, and semiclassical descriptions of the collision have been very fruitful in describing trap loss. The main difficulty in describing such collisions is in treating the strong excited-state spontaneous decay on the long time scale of the collision. We will see that a good, simple quantum mechanical description of such collisions is given by extending the above Condon point excitation picture to include excited-state decay during the collision. Quantitative theory is still made difficult by the presence of complex molecular hyperfine structure, except in certain special cases. Since the Condon points are at much smaller $R$ in the case of optical shielding, excited-state decay during the collision in not a major factor in the dynamics, and the primary challenge to theory in the case of optical shielding is how to treat the effect of a strong radiation field, which does strongly modify the collision dynamics. Simple semiclassical pictures seem not to work very well in accounting for the details of shielding, and a quantum description is needed. The theory of photoassociation is the most quantitative and well developed of all. Since the Condon point is typically inside the centrifugal barriers and $R_Q$, only the lowest few partial waves contribute to the spectrum. Although the line shapes of the free–bound transitions exhibit the strong influence of the quantum threshold laws, they can be quantitatively described by Franck–Condon theory, which can include ground- and excited-state hyperfine structure. We will show how photoassociation spectroscopy has become a valuable tool for precision spectroscopy and the determination of ground-state scattering lengths.

# 3

# Experimental methods of cold collisions

## 3.1 Atom traps

### 3.1.1 Light forces

It is well known that a light beam carries momentum and that the scattering of light by an object produces a force. This property of light was first demonstrated by Frish [139] through the observation of a very small transverse deflection ($3 \times 10^{-5}$ rad) in a sodium atomic beam exposed to light from a resonance lamp. With the invention of the laser, it became easier to observe effects of this kind because the strength of the force is greatly enhanced by the use of intense and highly directional light fields, as demonstrated by Ashkin [20, 21] with the manipulation of transparent dielectric spheres suspended in water. The results obtained by Frish and Ashkin raised the possibility of using light forces to control the motion of neutral atoms. Although the understanding of light forces acting on atoms was already established by the end of the 1970s, unambiguous demonstration of atom cooling and trapping was not accomplished before the mid 1980s. In this section we discuss some fundamental aspects of light forces and schemes employed to cool and trap neutral atoms.

The light force exerted on an atom can be of two types: a dissipative, spontaneous force and a conservative, dipole force. The spontaneous force arises from the impulse experienced by an atom when it absorbs or emits a quantum of photon momentum. When an atom scatters light, the resonant scattering cross section can be written as $\sigma_0 = \pi \lambda_0^2 / 2$, where $\lambda_0$ is the on-resonance wavelength, and the absorption is integrated over the natural line width. In the optical region of the electromagnetic spectrum the wavelengths of light are on the order of several hundreds of nanometers, so resonant scattering cross sections become quite large, $\sim 10^{-9}$ cm$^2$. Each photon absorbed transfers a quantum of momentum $\hbar k$ to the atom in the direction of propagation ($\hbar$ is the Planck constant divided by $2\pi$, and $k = 2\pi/\lambda$ is the magnitude of the wave vector associated with the optical field). The spontaneous emission following the absorption occurs in random directions; and, over many absorption–emission cycles, it averages to zero. As a result, the *net* spontaneous force acts on the atom in the direction of the light propagation, as shown schematically in Fig. 3.1. The saturated rate of photon scattering by spontaneous emission (half the reciprocal of the excited-state lifetime) fixes the upper limit to the force magnitude. This force is sometimes called *radiation pressure*.

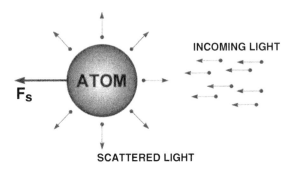

**Fig. 3.1.** Spontaneous emission following absorption occurs in random directions, but absorption from a light beam occurs along only one direction.

The dipole force can be readily understood by considering the light as a classical wave. It is simply the time-averaged force arising from the interaction of the dipole induced by the oscillating electric field of the light, with the gradient of the electric field amplitude. Focusing the light beam controls the magnitude of this gradient, and detuning the optical frequency below or above the atomic transition controls the sign of the force acting on the atom. Tuning the light below resonance attracts the atom to the center of the light beam while tuning above resonance repels it. The dipole force is a stimulated process in which no net exchange of energy between the field and the atom takes place, but photons are absorbed from one mode and reappear by stimulated emission in another. Momentum conservation requires that the change of photon propagation direction from initial to final mode imparts a net recoil to the atom. Unlike the spontaneous force, there is in principle no upper limit to the magnitude of the dipole force since it is a function only of the field gradient and detuning.

We can bring these qualitative remarks into focus by considering the amplitude, phase, and frequency of a classical field interacting with an atomic transition dipole. A detailed development of the following results is out of the scope of the present work, but can be found elsewhere [80, 362]. The usual approach is semiclassical and consists in treating the atom as a two-level quantum system and the radiation as a classical electromagnetic field [81]. A full quantum approach can also be employed [82, 156]; but it will not be discussed here.

The basic expression for the interaction *energy* is

$$U = -\boldsymbol{\mu} \cdot \mathbf{E}, \tag{3.1}$$

where $\boldsymbol{\mu}$ is the transition dipole and $\mathbf{E}$ is the electric field of the light. The *force* is then the negative of the spatial gradient of the potential,

$$\mathbf{F} = -\boldsymbol{\nabla}_R \cdot U = \mu \boldsymbol{\nabla}_R E, \tag{3.2}$$

where we have set $\boldsymbol{\nabla}_R \mu$ equal to zero because there is no spatial variation of the dipole over the length scale of the optical field. The optical-cycle average of the force is expressed as

$$\langle \mathbf{F} \rangle = \langle \mu \nabla_R E \rangle = \mu \{ (\nabla_R E_0) u + [E_0 \nabla_R (k_L R)] v \}, \tag{3.3}$$

where $u$ and $v$ arise from the steady-state solutions of the optical Bloch equations,

$$u = \frac{\Omega}{2} \cdot \frac{\Delta \omega_L}{(\Delta \omega_L)^2 + (\Gamma/2)^2 + \Omega^2/2} \tag{3.4}$$

and

$$v = \frac{\Omega}{2} \cdot \frac{\Gamma/2}{(\Delta \omega_L)^2 + (\Gamma/2)^2 + \Omega^2/2}. \tag{3.5}$$

In Eqs. 3.4 and 3.5 $\Delta \omega_L = \omega - \omega_0$ is the detuning of the optical field from the atomic transition frequency $\omega_0$, $\Gamma$ is the natural width of the atomic transition, and $\Omega$ is termed the Rabi frequency and reflects the strength of the coupling between field and atom,

$$\Omega = -\frac{\mu \cdot \mathbf{E}_0}{\hbar}. \tag{3.6}$$

In writing Eq. 3.3 we have made use of the fact that the time-average dipole has in-phase and in-quadrature components,

$$\langle \mu \rangle = 2\mu \left( u \cos \omega_L t - v \sin \omega_L t \right), \tag{3.7}$$

and the electric field of the light is given by the classical expression,

$$E = E_0 \left[ \cos \left( \omega t - k_L R \right) \right]. \tag{3.8}$$

The time-averaged force, Eq. 3.3, consists of two terms: the first term is proportional to the gradient of the electric field amplitude and the second term is proportional to the gradient of the phase. Substituting Eqs. 3.4 and 3.5 into Eq. 3.3, we have for the two terms,

$$\langle \mathbf{F} \rangle = \mu \left( \nabla_R E_0 \right) \cdot \frac{\Omega}{2} \cdot \left[ \frac{\Delta \omega_L}{(\Delta \omega_L)^2 + (\Gamma/2)^2 + \Omega^2/2} \right] \tag{3.9}$$
$$+ \mu [E_0 \nabla_R (-k_L R)] \cdot \frac{\Omega}{2} \cdot \left[ \frac{\Gamma/2}{(\Delta \omega_L)^2 + (\Gamma/2)^2 + \Omega^2/2} \right].$$

The first term is the dipole force, sometimes called the trapping force, $F_T$, because it is a conservative force and can be integrated to define a trapping potential for the atom:

$$\mathbf{F}_T = \mu \left( \nabla_R E_0 \right) \cdot \frac{\Omega}{2} \cdot \left[ \frac{\Delta \omega_L}{(\Delta \omega_L)^2 + (\Gamma/2)^2 + \Omega^2/2} \right], \tag{3.10}$$

and

$$U_T = -\int F_T \, dR = \frac{\hbar \Delta \omega_L}{2} \ln \left[ 1 + \frac{\Omega^2/2}{(\Delta \omega_L)^2 + (\Gamma/2)^2} \right]. \tag{3.11}$$

The second term is the spontaneous force, sometimes called the radiation pressure force or cooling force, $F_C$, because it is a dissipative force and can be used to cool atoms,

$$\mathbf{F}_C = -\mu E_0 \mathbf{k}_L \cdot \frac{\Omega}{2} \cdot \left[ \frac{\Gamma/2}{(\Delta \omega_L)^2 + (\Gamma/2)^2 + \Omega^2/2} \right]. \tag{3.12}$$

Note that in Eq. 3.10 the line-shape function in square brackets is dispersive and changes sign as the detuning $\Delta\omega_L$ changes sign from negative (red detuning) to positive (blue detuning). In Eq. 3.12 the line-shape function is absorptive, peaks at zero detuning, and exhibits a Lorentzian profile. These two equations can be recast to bring out more of their physical content. The dipole force can be expressed as

$$\mathbf{F}_T = -\frac{1}{2\Omega^2}\nabla\Omega^2\hbar\Delta\omega_L\left(\frac{S}{1+S}\right), \tag{3.13}$$

where $S$, the *saturation parameter*, is defined to be

$$S = \frac{\Omega^2/2}{(\Delta\omega_L)^2 + (\Gamma/2)^2}. \tag{3.14}$$

In Eq. 3.14 the saturation parameter essentially defines a criterion to compare the time required for stimulated and spontaneous processes. If $S \ll 1$ then spontaneous coupling of the atom to the vacuum modes of the field is fast compared to the stimulated Rabi oscillation, and the field is considered weak. If $S \gg 1$ then the Rabi oscillation is fast compared to spontaneous emission and the field is said to be strong. With the help of the definition of the Rabi frequency, Eq. 3.6, and the light beam intensity,

$$I = \frac{1}{2}\epsilon_0 c E_0^2. \tag{3.15}$$

Equation 3.13 can be written in terms of the gradient of the light intensity, the saturation parameter and the detuning,

$$\mathbf{F}_T = -\frac{1}{4I}(\nabla I)\hbar\Delta\omega_L\left(\frac{S}{1+S}\right). \tag{3.16}$$

Note that negative $\Delta\omega_L$ (red detuning) produces a force attracting the atom to the intensity maximum while positive $\Delta\omega_L$ (blue detuning) repels the atom away from the intensity maximum.

The spontaneous force or cooling force can also be written in terms of the saturation parameter and the spontaneous emission rate,

$$\mathbf{F}_C = \frac{\hbar\mathbf{k}_L\Gamma}{2}\left(\frac{S}{1+S}\right), \tag{3.17}$$

which shows that this force is simply the rate of absorption and re-emission of momentum quanta $\hbar k_L$ carried by a photon in the light beam. Note that as $S$ increases beyond unity,

$$\mathbf{F}_C \longrightarrow \frac{\hbar\mathbf{k}_L\Gamma}{2},$$

the photon scattering rate for a saturated transition.

### 3.1.2 The magneto-optical trap (MOT)

*Basic notions*

Pritchard *et al.* [318] originally suggested that the spontaneous light force could be used to trap neutral atoms. The basic concept exploited the internal degrees of freedom of the atom as a way of circumventing the optical Earnshaw theorem (OET) proved in [22]. This theorem states that if a force is proportional to the light intensity, its divergence must be null because the divergence of the Poynting vector, which expresses the directional flow of intensity, must be null through a volume without sources or sinks of radiation. This null divergence rules out the possibility of an inward restoring force everywhere on a closed surface. However, when the internal degrees of freedom of the atom are considered, they can change the proportionality between the force and the Poynting vector in a position-dependent way such that the OET does not apply. Spatial confinement is then possible with spontaneous light forces produced by counterpropagating optical beams. Using these ideas to circumvent the OET, Raab *et al.* [321] demonstrated a trap configuration that is presently the most commonly employed. It uses a magnetic field gradient produced by a quadrupole field and three pairs of circularly polarized, counterpropagating optical beams, detuned to the red of the atomic transition and intercepting at right angles in the position where the magnetic field is zero. The magneto-optical trap exploits the position-dependent Zeeman shifts of the electronic levels when the atom moves in the radially increasing magnetic field. The use of circularly polarized light, red-detuned by $\sim\Gamma$ results in a spatially dependent transition probability, the net effect of which is to produce a restoring force that pushes the atom toward the origin. To make clear how this trapping scheme works, consider a two-level atom with a $J = 0 \rightarrow J = 1$ transition moving along the $z$ direction (the same arguments will apply to the $x$ and $y$ directions). We apply a magnetic field $B(z)$ increasing linearly with distance from the origin. The Zeeman shifts of the electronic levels are position-dependent, as shown in Fig. 3.2(a). We also apply counterpropagating optical fields along the $\pm z$ directions carrying opposite circular polarizations and detuned to the red of the atomic transition. It is clear from Fig. 3.2 that an atom moving along $+z$ will scatter $\sigma^-$ photons at a faster rate than $\sigma^+$ photons because the Zeeman effect will shift the $\Delta M_J = -1$ transition closer to the light frequency.

$$F_{+z} = -\frac{\hbar k}{2}\Gamma\frac{\Omega^2/2}{\left(\Delta + kv_z + \frac{\mu_B}{\hbar}\cdot\frac{dB}{dz}\cdot z\right)^2 + (\Gamma/2)^2 + \Omega^2/2}. \tag{3.18}$$

In a similar way, if the atom moves along $-z$ it will scatter $\sigma^+$ photons at a faster rate from the $\Delta M_J = +1$ transition.

$$F_{-z} = +\frac{\hbar k}{2}\Gamma\frac{\Omega^2/2}{\left(\Delta - kv_z - \frac{\mu_B}{\hbar}\cdot\frac{dB}{dz}\cdot z\right)^2 + (\Gamma/2)^2 + \Omega^2/2}. \tag{3.19}$$

The atom will therefore experience a net restoring force pushing it back to the origin. If the light beams are red-detuned by $\sim\Gamma$, then the Doppler shift of the atomic motion will

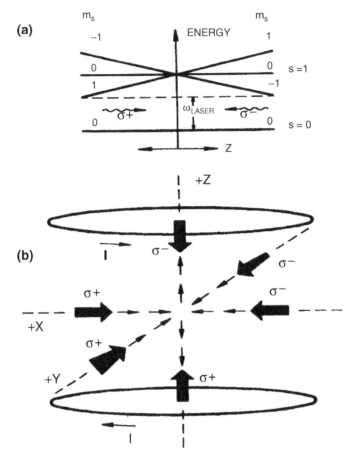

**Fig. 3.2.** MOT schema: (a) energy level diagram showing the shift of Zeeman levels as an atom moves away from the $z = 0$ axis. The atom encounters a restoring force in either direction from counterpropagating light beams. (b) A typical optical arrangement for the implementation of a magneto-optic trap.

introduce a velocity-dependent term to the restoring force such that, for small displacements and velocities, the total restoring force can be expressed as the sum of a term linear in velocity and a term linear in displacement,

$$F_{MOT} = F_{+z} + F_{-z} = -\alpha \, \dot{z} - K z. \tag{3.20}$$

Equation 3.20 expresses the equation of motion of a damped harmonic oscillator with mass $m$,

$$\ddot{z} + \frac{2\alpha}{m} \dot{z} + \frac{K}{m} z = 0. \tag{3.21}$$

The damping constant $\alpha$ and the spring constant $K$ can be written compactly in terms of the atomic and field parameters as

$$\alpha = \hbar k \Gamma \frac{16 \left| \Delta' \right| (\Omega')^2 (k/\Gamma)}{\left[1 + 2 (\Omega')^2\right]^2 \left[1 + \frac{4(\Delta')^2}{1+2(\Omega')^2}\right]^2} \tag{3.22}$$

and

$$K = \hbar k \Gamma \frac{16 \left| \Delta' \right| (\Omega')^2 \left(\frac{d\omega_0}{dz}\right)}{\left[1 + 2 (\Omega')^2\right]^2 \left[1 + \frac{4(\Delta')^2}{1+2(\Omega')^2}\right]^2}, \tag{3.23}$$

where $\Omega'$, $\Delta'$, and

$$d\omega_0/dz = \frac{\left[(\mu_B/\hbar)\left(\frac{dB}{dz}\right)\right]}{\Gamma}$$

are $\Gamma$ normalized analogs of the quantities defined earlier. Typical MOT operating conditions fix

$$\Omega' = \frac{1}{2} \qquad \Delta' = 1,$$

so $\alpha$ and $K$ reduce (in MKS units) to

$$\alpha(\mathrm{N\,m^{-1}\,s}) \simeq (0.132)\hbar k^2 \tag{3.24}$$

and

$$K(\mathrm{N\,m^{-1}}) \simeq (1.16 \times 10^{10})\hbar k \frac{dB}{dz}. \tag{3.25}$$

The extension of these results to three dimensions is straightforward if one takes into account that the quadrupole field gradient in the $z$ direction is twice the gradient in the $x$, $y$ directions, so that $K_z = 2K_x = 2K_y$. The velocity-dependent damping term implies that kinetic energy $E$ dissipates from the atom (or collection of atoms) as

$$E/E_0 = e^{-\frac{2\alpha}{m}t},$$

where $m$ is the atomic mass and $E_0$ is the kinetic energy at the beginning of the cooling process. Therefore the dissipative force term cools the collection of atoms as well as combining with the displacement term to confine them. The damping time constant

$$\tau = \frac{m}{2\alpha}$$

is typically tens of microseconds. It is important to bear in mind that a MOT is anisotropic since the restoring force along the $z$ axis of the quadrupole field is twice the restoring force in the $xy$ plane. Furthermore, a MOT provides a dissipative rather than a conservative trap, and it is therefore more accurate to characterize the maximum capture velocity rather than the trap "depth".

Early experiments with MOT-trapped atoms were carried out by slowing an atomic beam to load the trap, Raab *et al.* [321], Walker *et al.* [421]. Later a continuous uncooled source was used for that purpose, suggesting that the trap could be loaded with the slow atoms of a room-temperature vapor [66]. The next advance in the development of magneto-optical trapping was the introduction of the vapor-cell magneto-optical trap (VCMOT). As shown by Monroe *et al.* [292] this variation captures cold atoms directly from the low-velocity edge of the Maxwell–Boltzmann distribution always present in a cell background vapor. Without the need to load the MOT from an atomic beam, experimental apparatuses became simpler; and now many groups around the world use the VCMOT for applications ranging from precision spectroscopy to optical control of reactive collisions.

*Densities in a MOT*

The VCMOT typically captures about a million atoms in a volume less than a millimeter in diameter, resulting in densities of $\sim 10^{10}$ cm$^{-3}$. Two processes limit the density attainable in a MOT: (1) collisional trap loss and (2) repulsive forces between atoms caused by reabsorption of scattered photons from the interior of the trap (Walker *et al.* [421], Sesko *et al.* [346]). Collisional loss, in turn, arises from two sources: hot background atoms that knock cold atoms out of the MOT by elastic impact and binary encounters between the cold atoms themselves. Trap loss due to cold collisions is the topic of Chapter 4. The "photon-induced repulsion" or photon trapping arises when an atom near the MOT center spontaneously emits a photon that is reabsorbed by another atom before the photon can exit the MOT volume. This absorption results in an increase of $2\hbar k$ in the relative momentum of the atomic pair and produces a repulsive force proportional to the product of the cross sections for absorption of the incident light beam and for the scattered fluorescence. When this outward repulsive force balances the confining force, a further increase in the number of trapped atoms leads to larger atomic clouds, but not to higher densities.

### 3.1.3   Dark SPOT MOT

In order to overcome the "photon-induced repulsion" effect, Ketterle *et al.* [215] proposed a method that allows the atoms to be optically pumped to a "dark" hyperfine level of the atom ground state that does not interact with the trapping light. In a conventional MOT one usually employs an auxiliary "repumper" light beam, copropagating with the trapping beams but tuned to a neighboring transition between hyperfine levels of ground and excited states. The repumper recovers population that leaks out of the cycling transition between the two levels used to produce the MOT. As an example Fig. 3.3 shows the trapping and repumping transitions usually employed in an Na MOT. The scheme proposed by [215], known as a dark spontaneous-force optical trap (dark SPOT), passes the repumper through a glass plate with a small black dot shadowing the beam such that the atoms at the trap center are not coupled back to the cycling transition but spend most of their time ($\sim 99\%$) in

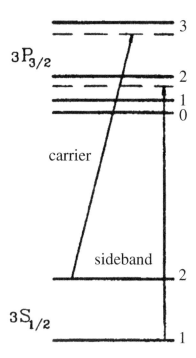

**Fig. 3.3.** Usual cooling (carrier) and repumping (sideband) transitions when optically cooling Na atoms. The repumper frequency is normally derived from the cooling transition frequency with electrooptic modulation. Dashed lines show that lasers are tuned about one natural line width to the red of the transition frequencies.

the "dark" hyperfine level. Cooling and confinement continue to function on the periphery of the MOT but the central core experiences no outward light pressure. The dark SPOT increases the density by almost two orders of magnitude.

### 3.1.4 The far-off resonance trap (FORT)

Although a MOT functions as a versatile and robust "reaction cell" for studying cold collisions, light frequencies must tune close to atomic transitions, and an appreciable steady-state fraction of the atoms remain excited. Excited-state trap-loss collisions and photon-induced repulsion limit achievable densities.

A far-off-resonance trap (FORT), in contrast, uses the dipole force rather than the sponta-neous force to confine atoms and can therefore operate far from resonance with a negligible population of excited states. The first atom confinement [73] was reported using a dipole-force trap. A hybrid arrangement in which the dipole force confined atoms radially while the spontaneous force cooled them axially was used by a NIST-Maryland collaboration (Gould *et al.* [158]) to study cold collisions, and a FORT was demonstrated by Miller *et al.* [279] for $^{85}$Rb atoms. The FORT consists of a single, linearly polarized, tightly focused Gaussian-mode beam tuned far to the red of resonance. The obvious advantage of large detunings is the suppression of photon absorption. Note from Eq. 3.12 that the spontaneous force, involving absorption and re-emission, falls off as the square of the detuning while Eq. 3.11 shows that the potential derived from dipole force falls off only as the detuning itself. At large detunings and high field gradients (tight focus) Eq. 3.11 becomes

$$U \simeq \frac{\hbar\Omega^2}{4\Delta\omega_L},$$                                    (3.26)

which shows that the potential becomes directly proportional to light intensity and inversely proportional to detuning. Therefore at far detuning but high intensity the depth of the FORT can be maintained. Most of the atoms will not absorb photons. The important advantages of FORTS compared to MOTS are: (1) high density ($\sim 10^{12}$ cm$^{-3}$) and (2) a well-defined polarization axis along which atoms can be aligned or oriented (spin polarized). The main disadvantage is the small number of trapped atoms due to small FORT volume. The best number achieved is about $10^4$ atoms [77, 78].

### Magnetic traps

Pure magnetic traps have also been used to study cold collisions, and they are critical for the study of dilute gas-phase Bose–Einstein condensates (BECs) in which collisions figure importantly. We anticipate therefore that magnetic traps will play an increasingly important role in future collision studies in and near BEC conditions.

The most important distinguishing feature of all magnetic traps is that they do not require light to provide atom containment. Light-free traps reduce the rate of atom heating by photon absorption to zero, an apparently necessary condition for the attainment of BEC. Magnetic traps rely on the interaction of atomic spin with variously shaped magnetic fields and gradients to contain atoms. Assuming the spin adiabatically follows the magnetic field, the two governing equations for the potential $U$ and force $\mathbf{F}_B$ are

$$U = -\boldsymbol{\mu}_S \cdot \mathbf{B} = -\frac{g_s\mu_B}{\hbar}\mathbf{S} \cdot \mathbf{B} = -\frac{g_s\mu_B}{\hbar}M_S B$$                          (3.27)

and

$$\mathbf{F}_B = -\frac{g_s\mu_B}{\hbar}M_S\boldsymbol{\nabla} B.$$                                    (3.28)

where $\mu_S$, $\mu_B$ are the electron spin dipole moment and Bohr magneton, respectively, $M_S$ the $z$-component of $\mathbf{S}$ and $g_s$ the Landé $g$-factor for the electron spin. If the atom has nonzero nuclear spin $\mathbf{I}$, then $\mathbf{F} = \mathbf{S} + \mathbf{I}$ substitutes for $\mathbf{S}$ in Eq. 3.27, the $g$-factor generalizes to

$$g_F \cong g_s\frac{F(F+1) + S(S+1) - I(I+1)}{2F(F+1)}$$                          (3.29)

and

$$\mathbf{F}_B = -\frac{g_F\mu_B}{\hbar}M_F\boldsymbol{\nabla} B.$$                                    (3.30)

Depending on the sign of $\mathbf{F}_B$ with respect to $\mathbf{B}$, atoms in states for which the energy increases or decreases with the magnetic field are called "weak-field seekers" or "strong-field seekers," respectively (for example, see Fig. 2.1). One could, in principle, trap atoms in any of these states, needing only to produce a minimum or a maximum in the magnetic field. Unfortunately only weak-field seekers can be trapped in a static magnetic field because such a field in free space can only have a minimum. Dynamic traps have been proposed to trap both weak- and strong-field seekers [248]. Even when weak-field seeking states are not in the lowest hyperfine levels they can still be used for trapping because the transition rate for spontaneous magnetic dipole emission is $\sim 10^{-10}\,\mathrm{s}^{-1}$. However, spin-changing collisions can limit the maximum attainable density.

The first static magnetic field trap for neutral atoms was demonstrated by Migdall *et al.* [278]. An anti-Helmholtz configuration (see Fig. 3.2), similar to a MOT, was used to produce an axially symmetric quadrupole magnetic field. Since this field design always has a central point of vanishing magnetic field, nonadiabatic Majorana transitions can take place as the atom passes through the zero point, transferring the population from a weak-field to a strong-field seeker and effectively ejecting the atom from the trap. This problem can be overcome by using a magnetic bottle with no point of zero field (see, for example, Gott *et al.* [157], Pritchard, [23, 316], Hess *et al.* [174]). The magnetic bottle, also called the Ioffe–Pritchard trap, was used to achieve BEC in a sample of Na atoms pre-cooled in a MOT (Mewes *et al.* [273]). Other approaches to eliminating the zero field point are the time-averaged orbiting potential (TOP) trap (Anderson *et al.* [15]) and an optical "plug" [103] that consists of a blue-detuned intense optical beam aligned along the magnetic trap symmetry axis and producing a repulsive potential to prevent atoms from entering the null field region. Trap technology continues to develop. Typically, under good conditions, $\sim 10^7$ atoms can be trapped in a Bose–Einstein condensate loaded from a dark spot MOT containing $\sim 10^9$ atoms. An alternative to magnetic traps, all-optical traps for BEC, have been demonstrated by Barrett *et al.* [37] and Grimm *et al.* [162, 163]. We will discuss optical traps in Chapter 7.

## 3.2  Atom beams

### 3.2.1  Introduction

Although MOTs and FORTs have proved productive and robust experimental tools for the investigation of ultracold collisions, they suffer from two shortcomings of any "reaction cell" technique: (1) collision kinetic energy or temperature cannot be varied over a wide range and (2) the isotropic distribution of molecular collision axes in the laboratory frame masks effects of light-field polarization on inelastic and reactive processes. Furthermore, probe beams in a MOT cannot explore close to the cooling transition without severely perturbing or destroying the MOT.

### 3.2.2  Velocity group selection

An alternative approach to atom traps is the study of collisions between and within atomic beams. In fact, the application of thermal atomic and molecular beams to investigate inelastic and reactive collisions dates from the early 1960s [170]. It was not until the early 1980s,

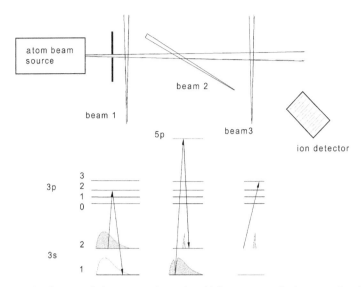

**Fig. 3.4.** Schematic of an atomic beam experiment in which a narrow velocity group is selected (laser beam 2) and intravelocity group collisions are probed (laser beam 3), from Thorsheim *et al.* [384].

however, that polarization and velocity-group selection led to detailed measurements of collision rate constants as a function of velocity, alignment, and orientation of the colliding partners [437]. At the start of the 1990s, Thorsheim *et al.* [384] reported the first measurements of subthermal collisions in an atomic beam. In this experiment a single-mode laser beam excited a narrow velocity group from a thermal sodium atomic beam, and the rate of associative ionization *within the selected velocity group* was measured as a function of light polarization perpendicular and parallel to the beam axis. The effective collision temperature of 65 mK was just on the threshold of the subthermal regime, defined as $T_S$, the temperature at which the atomic radiative lifetime becomes comparable to the collision duration. According to [202], for sodium excited to the 3p $^2P_{3/2}$, $T_S$ is 64 mK. Building on this experiment, Tsao *et al.* [403] used a much narrower optical transition (5p $^2P_{3/2}$) to select a velocity group from a collimated atomic beam and reported a rate constant for photoassociative ionization (PAI) at 5.3 mK, well within the subthermal regime. Figure 3.4 shows a diagram of the experimental set-up. Atoms emerging from a mechanically collimated thermal source interact with light in three stages: first, the atom beam crosses an optical "pump" beam tuned to transfer population from $F = 2$ to $F = 1$ of the ground state. Secondly, a velocity-group-selecting optical beam, tuned to the 3s$^2$ S$_{1/2}$ ($F = 1$) → 5p $^2P_{3/2}$ transition crosses the atomic beam at $\sim$45° and pumps a very narrow group to the previously emptied $F = 2$ level of the ground state. Finally, a third light beam, detuned by $\sim\Gamma$ to the red of the D$_2$ transition, excites collision pairs from within the $F = 2$ velocity group to form the detected PAI product ions. The production rate of the Na$_2^+$ product ions is then measured as a function of intensity and polarization. Although the optical-velocity-selection technique achieves subthermal collision temperatures, it suffers from the disadvantage that only a small fraction of the atoms in the atomic beam participate, and signal levels are consequently low.

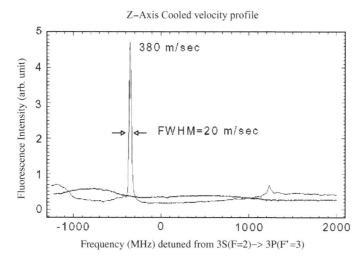

**Fig. 3.5.** The trace with the sharp peak is the velocity profile of axially cooled atomic beam. The broad low-amplitude trace shows the residual thermal velocity distribution from which the sharp peak was compressed, from Tsao *et al.* [405].

Tsao *et al.* [405] used this technique to investigate the polarization properties of optical suppression, an optically controlled collisional process described in Chapter 6.

### 3.2.3  Bright slow beams

Subjecting the axial or transverse velocity components of the beam to dissipative cooling dramatically compresses the phase space of the atom flux, resulting in dense, well-collimated atomic beams suitable for the study of atom optics, atom holography, or ultracold collision dynamics. Prodan *et al.* [319] first demonstrated the importance of this phase-space compression. In fact, atomic beams can now achieve a level of "brightness" (atom beam flux density per unit solid angle) many times greater than the phase-space conservation limit imposed by the Liouville theorem (*cf.* Pierce [313], Sheehy *et al.* [350], Kuyatt [233]). As discussed by Camposeo *et al.* [68] two classes of cold-beam sources can be identified: (1) those that start with a conventional hot vapor source and use optical cooling to "condition" the beam and (2) those that start from a cold-atom trap such as a MOT, extracting the cold-atom flux by breaking the symmetry of the trapping forces in some way. Using sources of type 1, several groups have realized "atom funnels" or "2-D MOTS" to cool and compress the transverse components of atom beams (Nellessen *et al.* [303]; Riis *et al.* [327]; Faultsich *et al.* [127]; Hoogerland, [180]; Yu *et al.* [451]; Tsao *et al.* [404]; DeGraffenreid *et al.* [106]). Others have used type 2 sources to obtain slow, well-collimated beams, (Myatt *et al.* [300]; Swanson, *et al.* [377]; Lu, *et al.* [249]; Kim *et al.* [218]). Figure 3.5 shows how a cooling light beam counterpropagating to a thermal atomic beam can compress the Maxwell–Boltzmann distribution into a relatively narrow range of velocities.

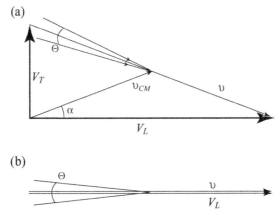

**Fig. 3.6.** Top panel: Newton diagram of a binary collision between two particles of equal mass. Note that the center-of-mass coordinate is located midway along the relative velocity vector. Laboratory longitudinal and transverse velocities are labelled $V_L$ and $V_T$, respectively. The molecular acceptance scattering is labelled $\Theta$. The center-of-mass velocity and angle are labelled $v_{CM}$ and $\alpha$, respectively. Bottom panel: a Newton diagram showing how, as $V_T \rightarrow 0$, the small $\Theta$ acceptance angle in the molecular frame aligns along the laboratory beam axis.

While the rate constants of cold and ultracold collisions can be measured in the "reaction bulb" environment of a magneto-optic trap, the isotropic distribution of collision axes limits these studies to spatially averaged quantities such as inelastic collision rate constants. However, the cold collision regime, where only a few partial waves participate, is just where alignment and orientation effects can be especially pronounced. A slow, highly collimated atom beam therefore provides an ideal environment for studying favorable classes of collision kinematics. In particular, inelastic or reactive processes proceeding through a two-step interaction sequence, the first occurring at long range and the second at short range, results in a very narrow acceptance angle in the entrance collision channel along the *molecular* axis of the approaching atoms. Examples of these kinds of collisions are charge transfer and harpooning reactive collisions [170]. If the laboratory divergence angle of the atom beam is also highly restricted, the molecular collision axis nearly superposes on the *laboratory* axis, providing a collision reference axis accessible to laboratory manipulation. Figure 3.6 shows the relevant Newton diagram for this kinematic case. We will see in Chapter 5 that photoassociative ionization falls into this category of two-step, long-range–short-range processes.

# 4

# Inelastic exoergic collisions in MOTs

An exoergic collision converts internal atomic energy to kinetic energy of the colliding species. When there is only one species in the trap (the usual case) this kinetic energy is equally divided between the two partners. If the net gain in kinetic energy exceeds the trapping potential or the ability of the trap to recapture, the atoms escape; and the exoergic collision leads to trap loss.

Of the several trapping possibilities described in Chapter 3, by far the most popular choice for collision studies has been the magneto-optical trap (MOT). A MOT uses spatially dependent resonant scattering to cool and confine atoms. If these atoms also absorb the trapping light at the initial stage of a binary collision and approach each other on an excited molecular potential, then during the time of approach the colliding partners can undergo a fine-structure-changing collision (FCC) or relax to the ground state by spontaneously emitting a photon. In either case electronic energy of the quasimolecule converts to nuclear kinetic energy. If both atoms are in their electronic ground states from the beginning to the end of the collision, only elastic and hyperfine changing collisions (HCC) can take place. Elastic collisions (identical scattering entrance and exit states) are not exoergic but figure importantly in the production of Bose–Einstein condensates (BEC). At the very lowest energies only s-waves contribute to the elastic scattering, and in this regime the collisional interaction is characterized by the scattering length. The sign and magnitude of the scattering length determine the properties of a weakly interacting Bose gas and controls the rate of evaporative cooling, needed to achieve BEC. The HCC collisions arise from ground-state splitting of the alkali atoms into hyperfine levels due to various orientations of the non zero nuclear spin. A transition from a higher to a lower molecular hyperfine level during the collisional encounter releases kinetic energy. In the absence of external light fields hyperfine-changing collisions often dominate trap heating and loss.

If the collision starts on the excited level, the long-range dipole–dipole interaction produces an interatomic potential varying as $\pm C_3/R^3$. The sign of the potential, attractive or repulsive, depends on the relative phase of the interacting dipoles. For the trap-loss processes that concern us in this chapter we concentrate on the attractive long-range potential, $-C_3/R^3$. Owing to the extremely low energy of the collision, this long-range potential acts on the atomic motion even when the pair are as far apart as $\lambda/2\pi$ (the inverse of the light field wave vector $k$). Since the collision time is comparable to or greater than the excited state life time, spontaneous emission can take place during the atomic encounter.

If spontaneous emission occurs, the quasimolecule emits a photon red-shifted from atomic resonance and relaxes to the ground electronic state with some continuum distribution of the nuclear kinetic energy. This conversion of internal electronic energy to external nuclear kinetic energy can result in a considerable increase in the nuclear motion. If the velocity is not too high, the dissipative environment of the MOT is enough to cool this radiative heating, allowing the atom to remain trapped. However, if the transferred kinetic energy is greater than the recapture ability of the MOT, the atoms escape the trap. This process constitutes an important trap loss mechanism termed radiative escape (RE), and was first pointed out by Vigué [415]. For alkalis there is also another exoergic process involving excited–ground collisions. Owing to the existence of fine structure in the excited state ($P_{3/2}$ and $P_{1/2}$), the atomic encounter can result in a fine-structure-changing collision, releasing $\Delta E_{FS}$ of kinetic energy, shared equally between both atoms. For example, in sodium

$$\frac{\Delta E_{FS}/k_B}{2} \simeq 12\,\text{K},$$

which can easily cause the escape of both atoms from the MOT, typically 1 K deep. These three effects, HCC, RE, and FCC, are the main exoergic collisional processes that take place in a MOT. They are the dominant loss mechanisms which usually limit the maximum attainable density and number in MOTs. They are not, however, the only type of collision in the trap, as we shall see later.

## 4.1   Excited-state trap loss theory

### 4.1.1   Early quasistatic models

*Gallagher–Pritchard model*

Soon after the first observations of ultracold collisions in traps, Gallagher and Pritchard [144] proposed a simple semiclassical model (GP model) to describe the FCC and RE trap loss processes. Figure 4.1 shows the two mechanisms involved. When atoms are relatively far apart they absorb a photon, promoting the system from the ground state to a long-range attractive molecular state. We denote the relevant molecular levels by their asymptotic energies. Thus, S + S represents two ground-state atoms, S + $P_{1/2}$ or S + $P_{3/2}$ represents one ground and one excited state atom, which can be in either one of the fine-structure levels. For radiative escape, the process is described by

$$A + A + \hbar\omega \rightarrow A_2^* \rightarrow A + A + \hbar\omega' \qquad (4.1)$$

with energy $\hbar(\omega - \omega')/2$ transferred to each atom. The fine-structure-changing collision is represented by

$$A + A + \hbar\omega \rightarrow A^*(P_{3/2}) + A \rightarrow A^*(P_{1/2}) + A + \Delta E_{FS} \qquad (4.2)$$

with $\Delta E_{FS}/2$ transferred to each atom. The GP model begins by considering a pair of atoms at the internuclear separation $R_0$. If this pair is illuminated by a laser of frequency $\omega_L$ and

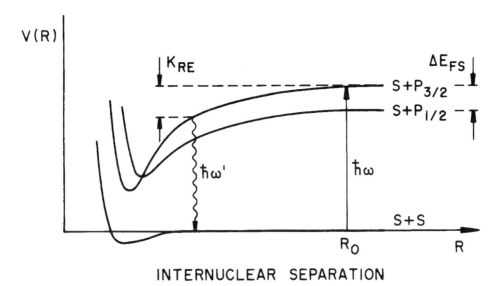

**Fig. 4.1.** Schematic diagram showing mechanisms of RE and FCC. An $\hbar\omega$ photon, red-detuned by $\sim\Gamma_A$, transfers population to the long-range attractive curve. Atoms accelerate radially along a molecular potential until either the molecule emits a photon $\hbar\omega'$ (RE) or undergoes a crossing to a potential curve asymptotically dissociating to ${}^2P_{1/2} + {}^2S_{1/2}$.

intensity $I$, the rate of molecular excitation, is given by

$$\Re(R_0, \omega_L, I) = \left[\frac{(\Gamma_M/2)^2}{(\Delta_M)^2 + (\Gamma_M/2)^2}\right] \cdot \frac{I}{\hbar\omega_L} \cdot \frac{\lambda^2}{2\pi} = \epsilon(\omega_L, R_0) \cdot \frac{I}{\hbar\omega_L} \cdot \frac{\lambda^2}{2\pi}, \qquad (4.3)$$

where $\Delta_M = [\omega_L - \omega(R_0)]$, $\omega(R_0) = \omega_A - C_3/\hbar R_0^3$ is the resonant frequency at $R_0$, and $\omega_A$ is the atomic resonant frequency. The constant $C_3$ characterizes the excited molecular potential, $\lambda^2/2\pi$ is the photoabsorption cross section to all attractive molecular states (half the atomic value), and $I$ and $\omega_L$ are the laser intensity and frequency, respectively. Finally, $\Gamma_M$ is the molecular spontaneous decay rate, here taken as twice the atomic rate, $\Gamma_A$.

After excitation at $R_0$, the atom pair begins to accelerate together on the $-C_3/R^3$ potential, reaching the short-range zone where the FCC and RE processes leading to trap loss occur. The time to reach the short-range zone is essentially the same as to reach $R = 0$. This time is easily calculated by integrating the equation of motion,

$$t(R_0) = \left(\frac{M}{4}\right)^{1/2} \int_0^{R_0} dR \left(\frac{C_3}{R^3} - \frac{C_3}{R_0^3}\right)^{-1/2} = 0.747 \left(\frac{M R_0^5}{4C_3}\right)^{1/2}, \qquad (4.4)$$

where $M$ is the atomic mass. Defining $\Delta = \omega_A - \omega_L$ as the detuning, $R_\tau$ as the interaction separation reached by the atoms, in one lifetime $\tau = \Gamma_M^{-1}$, and $\Delta_\tau$ the detuning at $R_\tau$. With these definitions, the time for the atoms to reach the short-range zone can be written as

$$t(R_0) = \left(\frac{\Delta_\tau}{\Delta}\right)^{5/6} \Gamma_M^{-1}, \tag{4.5}$$

and the probability of survival against spontaneous emission is given by $\gamma = \exp[-\Gamma_M t(R_0)]$ or $\gamma = \exp[(-\Delta_\tau/\Delta)^{5/6}]$. If the A*–A pair reaches the short-range region (where FCC take place) before radiating, it rapidly traverses this zone twice, yielding a probability for FCC equal $\eta_J = 2P(1 - P)$, where $P$ is the Landau–Zener single-transit curve-crossing probability. Repeated oscillations of the atoms through the crossing results in a probability $P_{FCC}$ that is the sum of the probabilities at each traversal,

$$P_{FCC} = \eta_J \gamma + \eta_J(1 - \eta_J)\gamma^3 + \cdots \tag{4.6}$$
$$= \frac{\eta_J \gamma}{(1 - \gamma^2 + \eta_J \gamma^2)}.$$

If $n$ is the total atomic density, the number of pairs with separation $R_0 \to R_0 + dR_0$ is $n^2 4\pi R_0^2 dR_0/2$, and the total rate of FCC per unit of volume is the integration over all internuclear separations,

$$R_{FCC} = \frac{n^2}{2} \int_0^\infty dR_0\, 4\pi R_0^2\, \Re(R_0, \omega_L, I) P_{FCC}(R_0). \tag{4.7}$$

The rate *constant* between two colliding ground state atoms is then

$$k_{FCC} = \frac{1}{2} \int_0^\infty dR_0\, 4\pi R_0^2\, \Re(R_0, \omega_L, I) P_{FCC}(R_0). \tag{4.8}$$

Alternatively the rate constant can be expressed in terms of a binary collision between one excited and one unexcited atom,

$$R_{FCC} = nn^* k'_{FCC}, \tag{4.9}$$

where $n^*$ is the excited atom density. At steady state, the rate of atom photo-excitation expressed as the product of photon flux, atom density, and absorption cross section (including line-shape function) equals the rate of fluorescence,

$$(I/\hbar\omega_L)n(\lambda^2/\pi)\left[\frac{\Gamma_A^2/4}{(\Delta_M)^2 + \Gamma_A^2/4}\right] = n^*\Gamma_A, \tag{4.10}$$

and combining Eqs. 4.3, 4.7, 4.9, and 4.10,

$$k'_{FCC} = \frac{1}{4}\left[\frac{(\Delta_M)^2 + \Gamma_A^2/4}{\Gamma_A/4} \int_0^\infty dR_0\, 4\pi R_0^2\, \epsilon(\omega_L, R_0) P_{FCC}(R_0)\right]. \tag{4.11}$$

The fine-structure change is not the only loss mechanism in the trap. Radiative escape is also important. The rate of RE, per unit of volume, can be obtained from Eqs. 4.8 or 4.11, when $P_{FCC}(R_0)$ is replaced by $P_{RE}(R_0)$, the probability of spontaneous emission during transit in an inner region, $R < R_E$. Here $R_E$ is defined as the internuclear separation at which the kinetic energy acquired by the colliding pair is enough to escape from the MOT. The probability $P_{RE}(R_0)$ can be obtained as a series sum similar to Eq. 4.6. Defining the time spent at $R < R_E$ as $2t_E(R_0)$ obtained from an integration similar to Eq. 4.5, one writes $P_{RE}(R_0)$ as

$$P_{RE}(R_0) = \frac{2t_E(R_0)\Gamma_M\gamma}{(1 - \gamma^2 + \eta_J\gamma^2)}. \tag{4.12}$$

The GP model predicts that FCC dominates over RE. Gallagher and Pritchard [144] estimate $\eta_J \sim 0.2$ for sodium collisions, resulting in a prediction of $P_{FCC}/P_{RE} \sim 20$ for Na-MOT trap loss.

### Julienne–Vigué model

Julienne and Vigué [206] proposed an elaboration, hereafter termed the JV model, which introduced the role of angular momentum and a thermal averaging procedure, employed realistic molecular states, and took into account the effects of retardation on molecular spontaneous emission decay rates. Like Gallagher and Pritchard, they neglected molecular hyperfine structure. The rate of FCC or RE transition per unit of volume is written as

$$\frac{\text{rate}}{\text{volume}} = Kn^2 = \frac{1}{2d_g^2}\frac{\pi v}{k^2}\sum_{l,\epsilon}(2l+1)P_{TL}(\epsilon,l)P_{ES}(R,\epsilon,l,\Delta,I)n^2, \tag{4.13}$$

where $n$ is the ground-state density, $v$ is the asymptotic velocity for the atoms when they are far apart, $k$ is the wave vector associated with the momentum of the reduced particle, $I$ is the light intensity, $d_g$ is the ground-state degeneracy, and the factor of $\frac{1}{2}$ accounts for homonuclear symmetry. The summation is performed over all contributing attractive excited states indexed by $\epsilon$. The term $P_{TL}(\epsilon,l)$ represents the probability that trap loss (either FCC or RE) occurs at small internuclear separation $R_{TL}$ after the atoms have drawn together. This $P_{TL}(\epsilon,l)$ factor was evaluated using different methods and was shown to be nearly independent of the asymptotic energy of the atoms over a wide range of temperature. The JV expression Eq. 4.13 corresponds to the GP expression Eq. 4.7. Note, however, that excitation, survival against spontaneous decay, and the collisional interaction are factored differently. In GP, $\Re(R_0, \omega_L, I)$, represents excitation and $P_{FCC}$ or $P_{RE}$ represents survival and collisional interaction together; whereas in JV $P_{ES}$ represents excitation and survival and $P_{TL}$ the probability of the collisional process itself. The excitation and survival probability, $P_{ES}(R,\epsilon,\Delta,I)$, represents the probability that an excited state $\varepsilon$ produced at a rate $G(R')$ at some $R' > R$ by light detuned $\Delta$ and intensity $I$ will survive without spontaneous radiative decay during the motion from $R'$ to $R$. In Julienne–Vigué theory, $P_{ES}$ can be written as

$$P_{ES}(R,\epsilon,\ell,\Lambda,I) = \int_R^\infty G(R',\epsilon,\Delta,I)S_{Ei}^\ell(R,R',\epsilon)\frac{dR'}{v_{Ei}^\ell(R')}, \tag{4.14}$$

where $S_{Ei}^\ell(R,R',\epsilon)$ is the survival factor for the atom pair moving from $R'$ to $R$ on the excited state $\epsilon$, with angular momentum $\ell$ and the initial kinetic energy $Ei$, starting from the ground state. The excitation *rate* is $G(R',\epsilon,\Delta,I)$ and the excitation probability is given by $G(R',\epsilon,\Delta,I)\cdot\frac{dR'}{v_{Ei}^\ell(R')}$.

Trap loss collision rates in alkali MOTs have been measured extensively, and the specific molecular mechanisms for FCC and RE trap loss for this class of atoms have been discussed

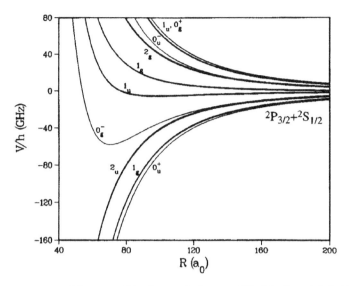

**Fig. 4.2.** Long-range potential curves of the first excited states of the Na dimer labeled by Hund's case (c) notation. The number gives the projection of the electron spin plus orbital angular momentum on the molecular axis. The g/u and +/− labels correspond to conventional point-group symmetry notation.

theoretically by Julienne and Vigué [206]. The molecular structure of the alkali metals are qualitatively similar, and the long range potentials are well known. The long-range potentials coming from the $P_{3/2} + S_{1/2}$ separated Na atoms are shown in Fig. 4.2 and give rise to five attractive states ($1_u$, $0_g^-$, $2_u$, $1_g$ and $0_u^+$). These states have different decay rates, which makes the survival effect very sensitive to the initial excitation. As to the collisional interaction itself, Julienne and Vigué [206] verified, using quantum scattering calculations, that the predictions of Dashsevskaya et al. [101] concerning FCC were qualitatively correct. These calculations show that FCC arises from two principal mechanisms: spin–orbit (radial) coupling at short range and Coriolis (angular) coupling at long range and that gerade states contribute very little to the FCC process. Therefore Julienne and Vigué conclude that of all the $^2P_{3/2} + ^2S_{1/2}$ entrance channel states, only $0_u^+$ and $2_u$ contribute significantly to the FCC cross section. Dashsevskaya [100] points out three basic pathways for FCC: (1) spin–orbit mixing of the $0_u^+$ components of the $A^1\Sigma_u$ and $b^3\Pi_u$, where they cross at short range, (2) Coriolis mixing of the $\Omega = 0, 1, 2$ components of $b^3\Pi_u$, and Coriolis coupling of $0_u^+$ and $1_u$ states at long range. The dominant mechanism for the light alkalis with relatively small spin–orbit terms (Li, Na, K) are the Coriolis couplings. For the heavy alkalis (Rb, Cs) the short-range spin–orbit mechanism prevails. Julienne and Vigué [206] calculated FCC and RE for all alkalis and obtained rate coefficients for trap loss combining FCC and RE as shown in Fig. 4.3. Because the spin–orbit interaction in Li is small, FCC does not lead to appreciable trap loss. It becomes important only if the trap depth is less than 250 mK. The RE contribution for Li is also small due to the fast decay rate of the excited states, preventing close approach of the two partners along the accelerating $-C_3/R^3$

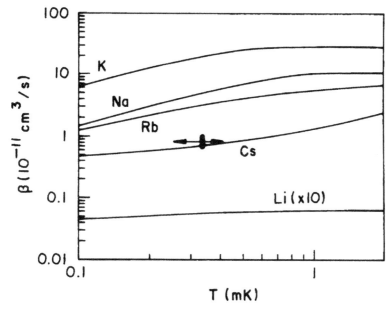

**Fig. 4.3.** Total trap loss rate constants $\beta$ as a function of temperature for the different alkalki species. These rates are calculated by Julienne and Vigué [206], and the solid points are measurements from Sesko *et al.* [345]. The arrow indicates the experimental uncertainty in temperature.

excited-state potentials. As a result, Li is predicted to have the lowest trap loss rate of all the alkalis.

### 4.1.2 Theoretical approaches to excited-state trap loss

*Quasistatic theories*

Both GP and JV theories develop rate expressions that derive from the "quasistatic" picture of light–matter interaction. The idea is that the atoms approach each other sufficiently slowly such that at each internuclear separation $R_c$ there is enough time to establish a steady state between rates of quasimolecule stimulated absorption and spontaneous emission. This steady state sets up the familiar Lorentzian line-shape function,

$$\frac{(\Gamma_M/2)^2}{\Delta_M^2 + (\Gamma_M/2)^2},$$

for the probability of excited-state population. This line-shape function permits off-resonant excitation around the Condon point $R_c$, where $\Delta_M = 0$. For a flat ground state $V_g = 0$ and excited state $V_e = -C_3/R^3$ the detuning is

$$\Delta = \frac{C_3}{\hbar R_c^3}$$

and

$$\left| \frac{d\Delta}{dR_c} \right| = \frac{3C_3}{\hbar R_c^4}.$$

Thus at relatively large detuning the line-shape function is still rather narrow and the excitation is fairly well confined around $R_c$. At small detunings, however, off-resonant excitation contributes significantly to the quasistatic rate constant, and further theoretical studies using quantum scattering with complex potentials [47, 48], Julienne *et al.* [205], optical Bloch equations, [32, 35], or quantum Monte Carlo methods; Lai *et al.* [235], Holland *et al.* [179] have shown that this off-resonant excitation picture is of dubious validity. These issues have been reviewed in detail by Suominen [370]. It is also worth noting that both GP and JV theories factor the rate constant into two terms: the first term describing excitation and survival at long range and the second term expressing the probability of the short-range trap-loss process itself. This decoupling of optical excitation and survival from the short-range collisional interaction provides a useful factorization even for theoretical approaches that do not invoke the quasistatic picture.

### 4.1.3   Method of complex potentials

At weak field, where optical cycling of population between ground and excited states can be safely ignored, a tractable quantum mechanical treatment of collisional loss processes can be conveniently introduced by insertion of an imaginary term in the time-independent Hamiltonian. Boeston *et al.* [48] and Julienne *et al.* [205] have each developed complex potential models of ultracold collisions in order to test the semiclassical theories. Both found serious problems with existing semiclassical theories at small detuning. Again the approach is to factor the probability for trap loss into two parts: (1) long-range excitation and survival, $J(E, \ell, \Delta, I; \Gamma)$ and (2) the short-range loss process itself, $P_X(E, \ell)$, where the subscript $X$ denotes FS or RE, and $E$ and $\ell$ are the total collision energy and angular momentum, respectively. The terms $\Delta$, $I$ represent the detuning and intensity of the optical excitation with $\Gamma$ being the spontaneous decay rate. The overall probability for the process is therefore,

$$P(E, \ell, \Delta, I; \Gamma) = P_X(E, \ell) \cdot J(E, l, \ell, \Delta, I; \Gamma). \tag{4.15}$$

In principle, the inclusion of a dissipative term such as spontaneous decay means that the problem should be set up using the optical Bloch equations (OBE) or quantum Monte Carlo methods, which take dissipation into account naturally over a wide range of field strengths. However, in the weak-field limit exact quantum scattering models with a complex potential suffice for testing the semiclassical GP and JV theories and developing alternative semiclassical models. Julienne *et al.* [207], for example, studied a three-state model comprising the ground-state entrance channel, an optically excited molecular state, and a fictitious probe state which simulates the collisional loss channel of the FS or RE processes. The trap-loss probability is then expressed in terms of the S-matrix connecting the ground state ($g$) to the probe state $(p)$,

$$P(E, \ell, \Delta, I; \Gamma) = |S_{gp}(E, \ell, \Delta, I; \Gamma)|^2 = P_{XQ}(E, \ell) \cdot J_Q(E, \ell, \Delta, I; \Gamma). \quad (4.16)$$

The subscript $Q$ denotes quantum close coupling, and Julienne *et al.* [207] compare $J_Q$ to four semiclassical approximations for $J$ : (1) $J_{JV}$ from the semiclassical JV model, (2) $J_{BJ}$ from an OBE calculation of [32], (3) $J_{BV}$ from the Landau–Zener (LZ) model of [47], and (4) $J_{LZ}$ from the LZ model of Julienne *et al.* [205]. Both $J_{BV}$ and $J_{LZ}$ are variants of the celebrated Landau–Zener formula,

$$J_{LZ} \simeq S_a e^{-A}, \quad (4.17)$$

where $S_a$ is the survival factor and $A$ is the familiar LZ argument,

$$A = \frac{2\pi |V_{ag}(I)|^2}{\hbar \left| \frac{dV_a}{dR} - \frac{dV_g}{dR} \right|_{R_c} v_a}. \quad (4.18)$$

In Eq. 4.18 $V_{ag}$ is the optical coupling, proportional to the laser intensity, and the denominator is the product of the difference in slopes evaluated at the Condon point $R_C$ and the local velocity $v_a$. The inner-region collisional interaction $P_{XQ}(E, \ell)$ is insensitive to the intensity and detuning of the outer-region optical excitation $\Delta, I$, the decay rate $\Gamma$, or the collision energy $E$. Therefore the overall trap-loss probability (Eq. 4.16) is controlled by the behavior of $J_Q(E, \ell, \Delta, I; \Gamma)$. Figure 4.4 shows Cs trap-loss probabilities for the various $J$'s as

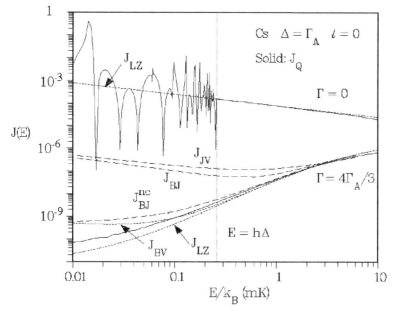

**Fig. 4.4.** Cs trap-loss probabilities for various $J$s as a function of the collision temperature. Notice the steep drop for $J_{BV}$, $J_{LZ}$, and $J_{BJ}^{nc}$. Notice also that semiclassical theory $J_{JV}$ completely misses this drop below about 1 mK, from Julienne *et al.* [207].

a function of collision temperature and reveals two important points. First, the quantal close coupling $J_Q$ confirms the conclusion of [48], which first pointed out a "quantum suppression" effect, i.e. that $J$ for a Cs model falls by about four orders of magnitude below the semiclassical prediction, $J_{BJ}$, as the temperature decreases from 1 mK to 10 μK. Secondly $J_{LZ}$ tracks the close-coupling results very closely even down into this "quantum suppression" region. At first glance the ability of the semiclassical Landau–Zener formula to follow the quantum scattering result even where the semiclassical OBE and JV theories fail may seem surprising. The very good agreement between $J_Q$ and $J_{LZ}$ simply means that the dramatic reduction in trap-loss probability below 1 mK can be interpreted essentially as poor survival against spontaneous emission when the approaching partners are locally excited at the Condon point at long range. The calculations in Fig. 4.4 were carried out with a detuning $\Delta$ of one atomic line width $\Gamma_A$. To check this interpretation, further calculations were carried out at greater red detunings and with lighter alkalis. In both cases the "quantum suppression" effect was greatly diminished as an increased survival interpretation would lead one to expect. An experimental measurement of the reduction in trap-loss rate with temperature has been reported by Wallace et al. [423] in $^{85}$Rb collisions.

### 4.1.4  Two-photon distorted wave theory

Solts et al. [357] have developed a distorted wave theory for evaluating weak-field, two-photon spectra, collision-induced energy redistribution, and radiative escape rates in cold atom traps. This theory complements the method of complex potentials discussed above and illustrates the connection between trap-loss spectra and photoassociation spectra. The two-photon distorted-wave (TPDW) collision theory is a rigorous quantum mechanical development, valid in the weak-field limit, and capable of treating multichannel collisional effects such as hyperfine structure. The theory concentrates on RE trap loss collisions,

$$A + A + \hbar\omega_L \rightarrow [A + A^*] \rightarrow A + A + \hbar\omega_S \tag{4.19}$$

treating the radiative coupling as weak perturbations, while fully taking into account the collision dynamics. The A term denotes an alkali atom or metastable rare gas atom, $[A + A^*]$ a bound excited quasimolecule, and $\omega_L, \omega_S$ are the exciting laser photon and re-emitted fluorescent photon, respectively. When $\omega_L$ is close to $\omega_0$, the A $\leftrightarrow$ A$^*$ atomic resonance frequency, so that the red-detuning $\Delta_L = \omega_L - \omega_0$ is within a few natural line widths of the atomic transition, the spectrum of $\hbar\omega_S$ will appear continuous and its asymmetrically peaked form is controlled by the pair distribution function dependence on $\Delta_L$ and on radiative damping. For an example of this line shape see the corresponding trap-loss spectrum in Fig. 4.24 in Section 4.2.2. As $|\Delta_L|$ increases, and the density of excited bound states decreases, the spectrum resolves into the discrete vibration–rotation progressions characteristic of photoassociation spectra. The state-to-state cross section for energy-specific $(E_i \rightarrow E_f)$ transitions is given by

$$\frac{\mathrm{d}\sigma(E_i, E_f)}{\mathrm{d}E_f} = \frac{2\pi^2}{k_i^2}\hbar\gamma \, |V_L|^2 \, Y(E_i, E_f), \tag{4.20}$$

where $k_i^2 = 2\mu E_i/\hbar^2$ and $V_L$ is the exciting laser coupling strength. The "characteristic strength" function $Y(E_i, E_f)$ can be expressed as a partial wave expansion,

$$Y(E_i, E_f) = \sum_{l=0}^{l_{max}} (2l+1) \left| \Theta_{fi}^{(l)} \right|^2,$$ (4.21)

where $\Theta_{fi}^{(l)}$ is the two-photon $T_{fi}$ matrix element $(E_f, l|T^{fi}|E_i, l)$ divided by the product of $\Delta_L$ and $\Delta_S = \omega_S - \omega_0$. The summation over partial waves is effectively cut off by a centrifugal barrier. The trap loss rate constant $\beta_{RE}(T)$ is calculated by averaging over the thermal distribution of atoms at the temperature of the trap and integrating over the final energies exceeding the trap escape threshold $\Delta_{RE}$,

$$\beta_{RE}(T) = \left( \frac{2\pi\hbar^2}{\mu k_B T} \right)^{3/2} \gamma \, |V_L|^2 \int_0^\infty dE_i \, I(\Delta_{RE}) e^{-E_i/k_B T},$$ (4.22)

where $k_B$ is the Boltzmann constant and

$$I(\Delta_{RE}) = \int_{\Delta_{RE}}^\infty dE_f \, Y(E_i, E_f).$$ (4.23)

At weak field the loss rate constant is directly proportional to the field intensity through the $|V_L|^2$ term. Equation 4.20 shows that the state-to-state redistribution cross section is proportional to the "characteristic strength function." Figure 4.5 shows the "spectrum" of $Y(E_i, E_f)$ as a function of detuning $\Delta_L$, taking the $0_u^+$ state of Na$_2$ as an example of an excited-state potential. The figure shows how the density of resonances decreases with

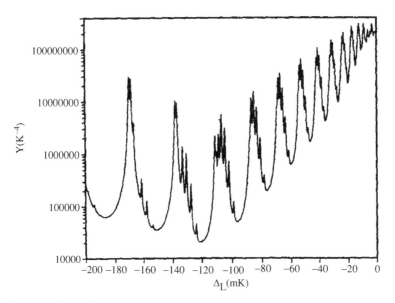

**Fig. 4.5.** Spectrum of the characteristic strength function, $Y$, as a function of detuning $\Delta_L$. The units of detuning in the figure are given in mK where 1 mK $\approx$ 20 MHz. Note that resonance peaks are resolvable even at small detuning due to the simplicity of the model calculation (Solts *et al.* [357]).

increasing detuning. These lines, typical of the photoassociation spectra to be discussed in Chapter 5, are clearly resolved even to very small detuning because only one spinless excited state is involved. If other excited states and hyperfine structure were included, the overlapping resonances would blend together into an effective continuum at small detuning, but would resolve into typical photoassociation spectra at large detuning (see, for example, Fig. 5.12.). In principle, the TPDW theory could be applied to a real photoassociation spectrum (at weak field) including hyperfine structure and excited states. This theory presents a unified view of trap-loss and photoassociation spectroscopy which shows them to be two aspects of the same cold-collision interaction with the laser field, differing only in the detuning regime.

## Optical Bloch equations

Both the GP and JV semiclassical models as well as the complex potential and TPDW quantum scattering calculations assume an initial optical excitation and subsequent spontaneous decay, after which the decayed population no longer participates in the collisional process. This picture is valid as a weak-field limit, but at higher fields the decayed population can be re-excited, resulting in an important modification of the overall probability of excitation at various points along the collision trajectory as well as changes in the ground-state kinetic energy. The need to treat dissipation at high field requires a density matrix approach; and, by introducing semiclassical trajectories, the time-dependent equations of motion governing the density matrix transform to the optical Bloch equations. Although the optical Bloch equations treat population recycling properly in principle, problems arise when the time dependence of these equations maps to a spatial dependence. This mapping converts the time coordinate to a semiclassical reference trajectory; but, because the potentials of the ground and excited states are quite different, the state vectors evolving on these two potentials, in a diabatic representation, actually follow very different trajectories. Reference [32] formulated two versions of an OBE trap-loss theory in a diabatic representation and calculated the excitation-survival factor, $J(E, \ell, \Delta, I; \gamma)$ of Eq. 4.15. One version applies correction factors to the initial reference trajectory so as to accurately reflect the actual semiclassical path followed by each state vector component. The other version makes no such corrections. Figure 4.4 shows that the "not-corrected" version, $J_{BJ}^{nc}$, actually works much better than the "corrected" version, $J_{BJ}$, although neither tracks the quantal results as well as $J_{LZ}$. The problem appears to be that the correction factors can successfully handle the diagonal terms of the density matrix, but fail when applied to the field-induced off-diagonal terms. These terms are important in a diabatic representation. Band et al. [35] have recast the OBE calculations in an adiabatic basis in which the field interaction is diagonalized and the interaction coherences are therefore eliminated. This approach yields much better results at low temperature, as can be seen in Fig. 4.6. A complete derivation of the adiabatic OBE method is given by Suominen, Band and coworkers [371], including interpretative LZ models adapted to the strong-field case. Although numerical results using the optical Bloch equations show reasonable agreement with experiments performed in Cs by Sesko et al. [345] and Na by Marcassa et al. [257], the quantum calculations of [47, 48], Julienne et al. [207], and Suonimen et al. [376] indicate that these results are fortuitous. A proper

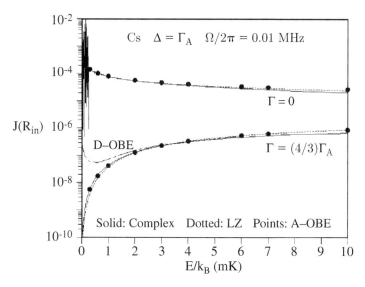

**Fig. 4.6.** Trap loss probability as a function of temperature for complex potential, Landau–Zener, and adiabatic OBE approaches. Notice that the adiabatic OBE is much better than the diabatic version, from [35].

quantum treatment using two-state models would give rate coefficients more than one order of magnitude smaller than measured.

### *Quantum Monte Carlo methods*

Suominen *et al.* [376] have carried out a quantum Monte Carlo two-state model study of survival in the very low-temperature regime in which $J_Q$, $J_{BV}$, $J_{LZ}$ of Fig. 4.4 all show a dramatic reduction in the trap-loss probability. The approach is to apply the Monte Carlo state vector method of [71] to the time evolution of wave packets undergoing random quantum jumps (Lai *et al.* [235]) during the course of the cold collision; and, by repeating the simulation many times, to gradually build up a statistical sample that approaches the density matrix. This approach treats spontaneous emission rigorously at all field strengths, taking into account not only population recycling but also ground-state kinetic energy redistribution. The motivation for the study was to provide definitive quantum mechanical benchmark calculations against which all approximate (but numerically more manageable) theories could be compared. Further motivation was to demonstrate, from an entirely different wave packet, time-dependent approach, the surprising resilience and robustness of the Landau–Zener formula in the very low-energy regime where a semiclassical expression would be normally suspect. Figure 4.7 shows the results of a diabatic OBE, a Monte Carlo wave packet, and a Landau–Zener calculation of the excited-state flux survival for Cs collisions at an energy of 100 μK. The diabatic OBE clearly fails, and the Landau–Zener result appears as the semiclassical average of the wave packet result. The conclusions of this benchmark study are therefore: (1) semiclassical, diabatic OBE calculations cannot be trusted below 1 mK; (2) the quantal wave packet calculations confirm previous findings that trap-loss collision probabilities drop off dramatically with temperature; and (3) the

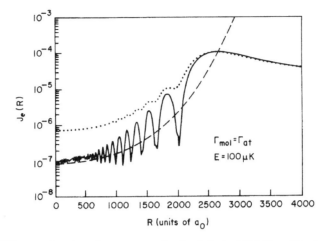

**Fig. 4.7.** Probability as a function of temperature for the diabatic OBE (dotted line), a wave packet (solid line), and Landau–Zener calculation (dashed line) of excited-state flux survival at a collision energy of 100 μK.

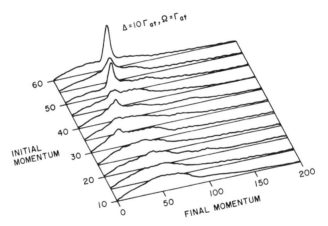

**Fig. 4.8.** Momentum distribution spreading due to re-excitation and population recycling, Holland *et al.* [179].

Landau–Zener results are in excellent accord with the oscillation-averaged fully quantal results.

Holland *et al.* [179] have applied the wave-packet quantum Monte Carlo technique to consider radiative heating of an ensemble of two-level Cs atoms; i.e. the reduced mass of the two-body collision was chosen to be that of Cs and the excited state was chosen to have the $C_3$ parameter of the attractive $0_u^+$ state. Figure 4.8 shows how the Monte Carlo method reveals momentum distribution spreading due to re-excitation and population recycling. Note that the spreading is especially marked for lower values of the initial momentum and results in ensemble heating even if trap loss does not occur. Significant heating without trap loss may have an important bearing on the limiting temperatures and densities attainable in optical traps.

### 4.1.5  Assessment of theoretical approaches

*Quasistatic vs dynamical*

The quasistatic theories of GP and JV assume that, for any pair separation $R_0$, the colliding partners are moving slowly enough that the field–quasimolecule coupling sets up a steady-state population distribution at $R_0$ governed by the Lorentzian of Eqs. 4.3 or 4.10. The characteristic time to establish this distribution is on the order of the spontaneous emission lifetime. Under the influence of an attractive $C_3/R^3$ potential the colliding nuclei rapidly pick up kinetic energy. Therefore at large detunings the time requirement to achieve the steady-state distribution is never fulfilled. At small detunings, where the nuclei accelerate more slowly, there *may* be sufficient time if the excitation is not too weak; but the governing Lorentzian distribution then implies very strong delocalization of excitation about the Condon point, $R_C$. This delocalization of excitation has been tested by quantum Monte Carlo wave packet studies (Suominen *et al.* [376]). The results demonstrate that excitation is a "dynamical" process, sharply localized around $R_C$ for all detunings, and established on a time scale of the order of the inverse of the excitation Rabi frequency. The quantum scattering complex potential studies and adiabatic OBE calculations confirm the dynamical picture, at least in the weak-field limit. Furthermore, the excitation-survival probability distribution itself can be factored into a survival term and an excitation term expressed by the Landau–Zener formula, $J_{LZ} \simeq S_a e^{-A}$, where $A$ is given by Eq. 4.18. Figure 4.4 shows that all dynamical theories correctly demonstrate the "quantum suppression" effect on trap loss probabilities as temperature descends from the mK to the μK regime, while the quasistatic steady-state theories completely miss it. At higher excitation fields, where theory must correctly take into account spontaneous emission, population recycling between ground and excited states, and kinetic energy redistribution on the ground state, so far only the quantum Monte Carlo wave packet studies of Suominen *et al.* [376] provide a reliable guide to the physics of collision processes. The calculational burden of these statistical samplings of the full density matrix has restricted these studies to two-level model systems. Studies of realistic potentials including hyperfine interaction are still beyond the scope of present calculational resources.

*Small vs large detuning*

The various theoretical approaches to trap loss at small detuning, i.e. $\Delta \leq 10\ \Gamma_A$ from atomic resonance, can at best provide a qualitative guide for understanding the process. No theory yet exists that is fully quantitative because small detuning implies an $R_C$ typically on the order of $\lambda/2\pi$, where $\lambda$ is the wavelength of the cooling transition. The large value of $R_C$ leads to two major complications: (1) the long time scale for the atoms to be accelerated to small $R$ where FCC or RE energy exchange occurs gives a prominent role to spontaneous decay and (2) the molecular hyperfine structure introduces great complexity into the long-range potential curves. All theoretical treatments to date ignore (2), because of the computational intractability of the full problem. It is beyond the scope of any conceivable numerical computational solution at the present. A full and proper treatment of the small detuning case, requires the solution to the von Neumann equation of motion for the quantum density matrix $\rho_{ij}(R, R')$ (Julienne, et al. [204]; Suominen et al. [376]). This is now feasible with Monte Carlo simulations for two-state models, as demonstrated by Suominen et al. [376], whose calculations provide the benchmarks against which any other two-state theories must be tested.

Unfortunately, all current theories of trap loss for small detuning suffer from the limitation of being restricted to two (or at least a very few) scattering channels. There are some general conclusions that can be extracted from these calculations, however. The first is that of the great usefulness of the semiclassical picture of localized excitation at the Condon point. As discussed in Section 4.1.5 above, the local equilibrium models of GP and JV give simple pictures for interpreting trap loss. As detuning changes from small (or the order of a few natural atomic widths) to large (many natural widths), these theories all show that the trap loss rate first increases with the magnitude of the detuning (as survival improves due to excitation at smaller $R$) and then decreases (survival approaches unity, but the available phase space of pairs proportional to $R^2$ decreases). Unfortunately, all of the quantum calculations show that in the small detuning regime the quasistatic picture with delocalized, Lorentzian excitation is fundamentally incorrect. The correct picture shows localized excitation at the Condon point followed by spontaneous decay as the atoms are accelerated together on the excited state. The Monte Carlo calculations even demonstrate that such a picture can even be extended to the strong-field saturation regime of two-state models: a semiclassical delayed decay model (Suominen et al. [371]) works where decay is only calculated from the inner edge of a region of saturation about the Condon point.

In contrast, at large detuning the GP and JV theories do result in excitation localized near the Condon point because of the steep slope of the upper potential curve, and in fact [32] showed that the GP theory becomes equivalent to the LZ + survival model which has now been demonstrated by the quantum calculations. In practice, this will be true for detunings on the order of 10 atomic line widths or more. The GP theory is especially useful in this regime. The validity of the simple GP formula has been verified in the trap loss studies to the red of the $S_{1/2} + P_{1/2}$ limit by Peters et al. [311]. Unfortunately, the semiclassical theories leave out the role of bound states of the excited potential, which are now well known from the detailed spectra measured by photoassociation spectroscopy at sufficiently large detuning, typically more than 100 line widths from atomic resonance. There is a need for the development of a theory that bridges the gap between the large detuning photoassociation region and the intermediate detuning regime where the density of bound vibrational states is so high that the quasicontinuum approach of GP applies. This bridging could be done by applying the complex potential method, which uses well-understood scattering algorithms, or the perturbative Franck–Condon approach of Solts et al. [357].

Another main complication at small detuning is the extreme complexity of the molecular hyperfine structure (Walker and Pritchard [420], Julienne et al., [207]; Lett et al. [245]). A number of the experimental studies described below indicate a prominent role for hyperfine structure in determining the trap loss rate. For example Fig. 4.9 shows the molecular hyperfine potentials originating from $^2S + ^2P_{3/2}$ Na atoms. The recent work of Fioretti

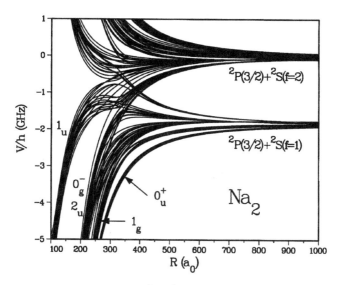

**Fig. 4.9.** Hyperfine molecular states near the $^2S + ^2P_{3/2}$ asymptote. The dense number of states gives rise to a large number of avoided crossings, state mixing, and congested spectra, from Lett *et al.* [245].

*et al.* [135] suggests that hyperfine optical pumping in a molecular environment during a collision may also be significant (see also Julienne and Mies [202]). The fundamental difficulty with the two-state treatments in the older GP, JV and OBE studies, is that the apparent agreement with experiment is only fortuitous. According to the quantum calculations, with their dynamical picture of the excitation, the predicted trap loss rates will be much lower (an order of magnitude or more) for these same two-state models than rates given by the older semiclassical theories. Therefore there is a fundamental discrepancy that remains to be understood. There is considerable likelihood that the resolution of this problem will require understanding the role of the complex molecular hyperfine structure, with molecular optical pumping and spontaneous decay. Although the semiclassical optical Bloch equation methods showed clear problems in the diabatic formulation, there is some promise in an adiabatic formulation of the semiclassical OBE method which appears much more accurate than the diabatic formulation (Suominen *et al.* [371]). Since it is unlikely that a fully quantum density matrix treatment including hyperfine structure will be available in the foreseeable future, a possible approach to understanding small-detuning trap loss may lie with adiabatic OBE or LZ semiclassical models ([47], Julienne *et al.* [207], Suominen *et al.* [371]).

The situation is much better at somewhat larger detuning, i.e. large compared to the atomic spontaneous decay line width. When the detuning becomes sufficiently large, the trap loss process gives rise to a discrete spectrum of trap loss due to the bound states in the excited attractive molecular potentials. The Condon point is sufficiently small that many vibrational cycles are necessary before spontaneous decay can occur, and the spacing between vibrational levels becomes large compared to their natural line width. We will discuss this case in detail in Chapter 5. In this limit of large detuning, the theory is in excellent shape, because spontaneous emission can be treated as a perturbation on the normal conservative multichannel scattering theory, and fully quantitative calculations including all molecular hyperfine structure have been carried out in great detail, and with superb agreement with experiment.

It is within the realm of possible calculations to study a realistic system without hyperfine structure. Group II species such as Mg, Ca, or Sr serve as examples. All of these have $^1S \rightarrow {}^1P$ transitions. We will show later in Chapter 6 on optical shielding how a full three-dimensional scattering treatment taking into account the degeneracy of the excited P level is possible. Monte Carlo simulations, as well as simpler models, of such a system appear to be feasible with advanced computation methods. Experimental studies of such a simple system would be very helpful in unravelling the complexities of small detuning trap loss. There is a realistic hope of testing proper theory against experimental measurements in such systems.

## 4.2  Excited-state trap-loss measurements

In general, collisions in traps are investigated by either detecting the product resulting from the binary encounter or through loss of the colliding species from the trap. When an ultracold collision releases kinetic energy, the reaction rate can be measured by observing the time dependence of the number of atoms $N$ either as the trap loads or as it decays. The equation governing the net loading rate of the trap is given by

$$\frac{\mathrm{d}N}{\mathrm{d}t} = L - \gamma N - \beta \int_v n^2(r, t)\, \mathrm{d}^3 r, \tag{4.24}$$

where $L$ is the capture rate, $\gamma$ is the rate constant for collisions of trapped atoms with a thermal background gas, and $\beta$ is the loss rate constant due to collisions between trapped atoms. In general, spatial variation of the atomic density $n(r, t)$ requires integration over the whole volume occupied by the atoms; but the spatial density distribution of the trapped atoms can be in one of two possible limits: at low density radiation trapping is negligible, and the trap spatial distribution is close to being Gaussian; in the high-density limit Walker *et al.* [421] showed that radiation trapping dominates and produces a constant spatial atom density. Thus, in the low-density limit the distribution $n(r, t)$ can be expressed as,

$$n(r, t) = n_0(t)e^{-r^2/w^2}$$

and Eq. 4.24 becomes

$$\frac{\mathrm{d}N}{\mathrm{d}t} = L - \gamma N - \frac{\beta N^2}{(2\pi)^{3/2} w^3}. \tag{4.25}$$

This equation has a complicated solution as shown by Prentiss *et al.* [315]; but $N(t)$ can be measured directly and $\beta$ determined by curve fitting. In the high-density limit, trap loading proceeds at constant density. As the number of atoms in the trap increases, the volume

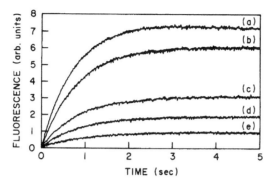

**Fig. 4.10.** Examples of trap loading rates in MOT (type-I trap with trapping laser tuned approximately one line width to the red of 2S ($F = 2$) → $3P_{3/2}(F' = 3$) for several different MOT intensities: (a) 96 mW cm$^{-2}$, (b) 70 mW cm$^{-2}$, (c) 36 mW cm$^{-2}$, (d) 23 mW cm$^{-2}$, and (e) 16 mW cm$^{-2}$. Curves are from Marcassa *et al.* [257].

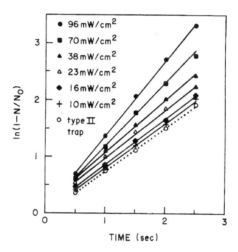

**Fig. 4.11.** Semilog plot of the integrated loading curves: absolute value of $1 - N/N_0$ versus $t$. The slope, $-(\gamma + \beta n_c)$, determines the trap loss rate constant $\beta$. Results are for a type-I trap with trapping laser tuned approximately one line width to the red of 2S ($F = 2$) → $3P_{3/2}(F' = 3$).

expands so that the density in the trap remains unchanged, and Eq. 4.24 takes the form,

$$\frac{dN}{dt} = L - (\gamma + \beta n_c) N, \tag{4.26}$$

where $n_c$ is the constant atomic density obtained in the trap. The value of $n_c$ is characteristic of the trapping conditions. The factor $\gamma + \beta n_c$ is determined by beginning with an empty trap, turning it on, and measuring the number of trapped atoms as a function of time. Equation 4.26 is then fitted to the measurement. Examples of these $N(t)$ transients obtained in a sodium MOT by Marcassa *et al.* [257] are presented in Fig. 4.10, at several different MOT laser intensities. A semilog plot of the absolute value of $1 - N/N_0$ versus $t$ determines the slope equal to $-(\gamma + \beta n_c)$ as shown in Fig. 4.11 The linearity of these plots confirms

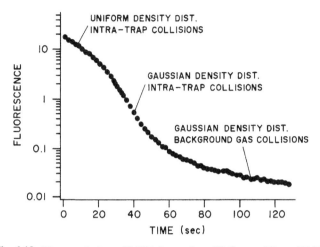

**Fig. 4.12.** Time evolution of MOT decay, from Walker and Feng [419].

that trap loading takes place within the regime of constant density, $n_c$. These loading-curve fits ignore the very early times where the constant density regime has not yet been reached. The density $n_c$ is obtained by first measuring the trap diameter with, for example, a charge-coupled-device (CCD) camera to determine the volume. Then fluorescence detection or absorption with an independent probe laser determines the number of trapped atoms. To extract $\beta$, the two-body trap-loss rate constant, it is necessary to devise an independent measurement of $\gamma$. This rate can be determined either by observing variations of $(\gamma + \beta n_c)$ for different background pressures of the trapped gas or by operating the MOT with a very low number of trapped atoms such that the term $\beta n_c$ is negligible compared with $\gamma$. Figure 4.11 shows the light-field-intensity dependence of $-(\gamma + \beta n_c)$, which is associated with the intensity dependence of $\beta$ since $\gamma$ and $n_c$ are insensitive to this parameter.

If the MOT is loaded from a slow atomic beam, the background loss rate $\gamma$ is normally very small and a convenient way to investigate $\beta$ is to observe the decay of the trapped atoms as a function of time after the loading process has been completed and the source of atoms interrupted. In this case the capture rate $L$ is no longer present in Eq. 4.24. Typical decay of a cesium MOT atomic fluorescence (Walker and Feng [419] ) as a function of time is shown in Fig. 4.12. In the initial stage, Fig. 4.12 shows a constant-density collision rate resulting in exponential decay. As the number of atoms decreases, the radiation trapping effect becomes weak enough that the density begins to drop while the trapped-atom distribution retains a Gaussian shape. In this regime the decay is nonexponential. Finally, when the number of atoms become very small, the intratrap collisions are no longer important and the collisions with the background gas cause the decay to once again become exponential in time, with a rate smaller than the first stage.

The trap-loss parameter $\beta$ contains the probabilities for inelastic processes such as fine-structure-changing collisions, radiative escape, and photoassociation. To better understand the influence of applied optical fields on these processes, several research groups have performed studies of the dependence of $\beta$ on various parameters, especially MOT laser

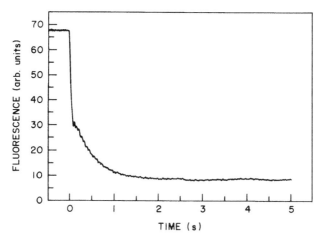

**Fig. 4.13.** Fluorescence–time spectrum. The abrupt drop at the initial moment is due to a change in light intensity without a variation in the number of trapped atoms, from Santos *et al.* [340]

frequency and intensity. Both the GP and JV models predict a $\beta$ increasing linearly with MOT intensity; initial experiments sought to verify this behavior.

Investigation of the variation of $\beta$ with MOT light intensity is straightforward to carry out. One can either load or unload the trap at different intensities and extract $\beta$ from the transient behavior. If the experiment is performed by loading a MOT from a slow atomic beam, this technique works well over a wide range of intensities (Sesko *et al.* [345], Wallace *et al.* [422], Kawanaka *et al.* [213], Ritchie *et al.* [329]). However, for vapor-cell loading where the $\gamma$ term in Eq. 4.24 becomes significant, the steady-state number of atoms in the trap at low MOT intensity becomes too small for reliable measurements. In this case, the study of $\beta$ as a function of intensity can be implemented by a sudden change of intensities as demonstrated by Santos *et al.* [340]. The basic idea is to switch between two different intensities in a short time interval. A large number of atoms are loaded into the trap at high intensity after which the intensity is suddenly attenuated by introducing a calibrated neutral density filter into the laser beam. At lower intensity the new loading and loss rates determine a new (lower) steady-state number of atoms. The transient variation of the number of trapped atoms between the two limits determines $\beta$ in the low-intensity regime. Figure 4.13 shows a typical fluorescence–time spectrum using a sudden change of intensity, starting from a high steady-state number of atoms. The fast drop observed at the first instant of time corresponds to the decrease in the photon scattering rate due to the change in light intensity without variation in the number of trapped atoms. The subsequent slow decay must be analyzed using Eq. 4.25 or Eq. 4.26 already described.

The investigation of $\beta$ with MOT detuning, is more difficult because the trap only operates over a narrow range of detunings $\sim\Gamma$ to the red of the atomic resonance. Hoffmann *et al.* [177] overcame this limitation by introducing an extra probe laser which they called a "catalysis" laser. This idea was first introduced by Sesko *et al.* [345]. Although the added optical field does not act as a catalyst in the conventional chemical sense of increasing reaction rates without a net consumption of the catalysis photon "reagent", the term has been widely adopted anyway. As long as the frequency of this laser is not too close to the MOT or repumping transition, it does not affect the trapping and cooling processes; but it can produce changes in the trap loss rate. Hoffmann *et al.* [177] assumed the effect of the catalysis laser on the trap loss rate to be additive, $\beta = \beta_t + \beta_c$, where $\beta_t$ represents the rate constant with only the trapping laser present and $\beta_c$ represents the contribution of the catalysis laser. It is important to note, however, that Sanchez-Villicana *et al.* [336] have demonstrated a "flux enhancement effect" (see the discussion later in Section 4.2.2 on rubidium trap loss and Fig. 4.25) that call the $\beta$ additivity assumption into question. Detuning the catalysis laser ($\Delta_c$) affects $\beta$ and the number of atoms in the trap, $N$,

$$N = \frac{L}{\gamma + \beta n_c f}, \tag{4.27}$$

where $L, \gamma$ have their usual meaning (see Eq. 4.24) and

$$f = \frac{\int n^2(\mathbf{r}, t) \, \mathrm{d}^3 \mathbf{r}}{n_c N}$$

is a measure of the deviation of the density distribution from the uniform density $n_c$. As the intensity $I_c$ and detuning $\Delta_c$ of the catalysis laser vary, the trap loss constant $\beta$ varies as,

$$\beta = \beta_t + \beta_c(\Delta_c) \frac{I_c(\Delta_c)}{I_{ref}}, \tag{4.28}$$

where $I_{ref}$ is a reference laser intensity taken to be $10 \, \mathrm{mW \, cm^{-2}}$. Equation 4.27 shows that varying $I_c$ and $\Delta_c$ in turn will alter $N$ and consequently the trap density, $n$. Unfortunately, the most difficult quantity to determine accurately in these experiments is the trap density, but Hoffmann *et al.* [177] avoided this problem by adjusting $I_c$ at each $\Delta_c$ such that $N$ and $n$ remained constant over the entire range of detunings in the experiment. With the condition that the product $\beta_c(\Delta_c)I_c(\Delta_c)$ be held invariant to maintain constant $N$, it was only necessary to measure the loading curve, Eq. 4.24, with and without the catalysis laser at one fixed detuning and intensity in order to obtain the absolute value of the $\beta_c I_c$ product. Once known, measures of $I_c(\Delta_c)$ determined absolute values of $\beta_c(\Delta_c)$ over the entire detuning range of the experiment. The reliability of the catalysis laser method depends strongly on the assumption that the extra laser light affects only the atomic collisions and not the performance and characteristics of the MOT. This assumption is best justified if the catalysis laser is not tuned too close to the trapping and repumping transitions and the power is not so high that light shifts begin to affect the MOT functioning.

Using these techniques to study trap loss, several specific cases have been investigated.

### Sodium trap loss

Prentiss *et al.* [315] carried out the first measurements of collisional losses in a MOT loaded from a cold sodium beam. In this work sudden interruption of the atomic beam stopped the loading at a well-defined instant of time and the subsequent fluorescence decay was recorded. At early times a nonexponential fluorescence decay was observed at a trap

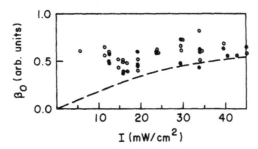

**Fig. 4.14.** Early experiment, Prentiss *et al.* [315], measuring the variation of trap loss rate constant in an Na MOT as a function of MOT intensity. The dashed line traces the calculated variation of the excited-state population.

density $\sim 7 \times 10^9$ cm$^{-3}$. Fitting of the nonexponential decay with an equation like Eq. 4.25 permitted extraction of a value for $\beta$. Investigation of the variation of $\beta$ with laser intensity is shown in Fig. 4.14. The data yielded a $\beta$ invariant with intensity (within $\pm 40\%$) over a range of 5–50 mW cm$^{-2}$. Variations of the trap depth, realized through changes in the MOT magnetic field, indicated that $\beta$ was insensitive to trap-depth changes as well. The absolute values of $\beta$ were measured to be about $4 \times 10^{-11}$ cm$^3$ s$^{-1}$ (within a factor of 5). Although Prentiss *et al.* [315] could not readily explain the unexpected insensitivity of $\beta$ to light intensity, they offered several possibilities for the observation including the effect of $m_F$ hyperfine levels mixing and shifts of the atomic energy levels that occur during a collision. The surprising lack of intensity dependence in their results underscored the importance of investigating binary collisions in MOTs.

Marcassa *et al.* [257] investigated $\beta$ over a wider range of intensity, from 20 to 300 mW cm$^{-2}$, using as Na vapor-cell-loaded MOT. In their report a definite light intensity dependence was indeed observed, and the apparent independence previously reported by Prentiss *et al.* [315] was explained by the relatively narrow range over which the earlier experiments had been carried out. In the experiment of Marcassa *et al.* [257] the MOT laser was tuned about 10 MHz to the red of the atomic cooling transition, and $\beta$ extracted from the fluorescence growth curve recorded during MOT loading, starting from an empty trap. Densities in the MOT ranged from $5 \times 10^9$ to $2 \times 10^{10}$ cm$^{-3}$, which is in the high-density limit where Eq. 4.26 governs the loading process. In order to distinguish $\gamma$ from $\beta n_c$ trap loss was measured using two tuning conditions, the type I and type II traps, described in Prentiss *et al.* [315]. The type II trap was operated in a density regime low enough so that trapped-atom collisions could be ignored and the slope of the fluorescence growth exponential determined $\gamma$. A calibrated telescope measured the MOT diameter from which the assumed spherical volume was calculated, and the atom number in the MOT was determined from fluorescence imaged onto a calibrated photomultiplier detection train. These two measurements determined $n_c$, and $\beta$ was then calculated from Eq. 4.26. The measured intensity dependence of $\beta$ is shown in Fig. 4.15. Despite error bars of 50% or more, due essentially to uncertainties in the volume measurement, the expected increase of $\beta$ with MOT intensity is unmistakable.

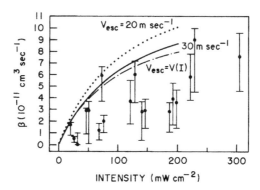

**Fig. 4.15.** Measurement of trap loss rate constant as a function of MOT light intensity. The wider range of light intensity reveals an increase in $\beta$ with MOT intensity. Dotted, full, and dot-dash curves are theory calculations using the OBE method of [32] with different assumptions concerning the maximum escape velocity of the MOT. The curve $V_{esc} = V(I)$ results from a simple model in which the escape velocity is a function of the MOT intensity. Data from Marcassa *et al.* [257].

Reference [32] proposed a new theory for trap loss, explicitly avoiding the local equilibrium assumption of the GP and JV models, and based on the optical Bloch equations. The OBE density-matrix approach to trap loss is appropriate to studies of optical field dependence because it naturally takes into account population recycling (optical excitation, decay, and re-excitation) at high field. In contrast, GP and JV are weak-field theories in which spontaneous emission is treated as a simple loss term, and the decayed population never recycles. Marcassa *et al.* [257] applied this OBE theory to their intensity dependence studies of Na trap loss by calculating $\beta$ as a function of MOT intensity and comparing it to the measured values. A difficulty in making this comparison is knowledge of the minimum velocity which must be imparted to the atoms in order to escape from the MOT. Assuming single-atom escape velocities of 20 and 30 ms$^{-1}$, Marcassa *et al.* [257] calculated an RE contribution to the total loss rate of 38% and 24%, respectively, with the remaining fraction being due to FCC. Figure 4.15 plots the results from calculation and measurement. Although the model did not include hyperfine structure of the excited states, the agreement between theory and experiment is reasonably satisfactory, with theory being somewhat higher than experiment. However, the assessment in Section 4.1.5 indicates that this agreement is likely to be fortuitous. It is worthwhile noting that Marcassa *et al.* [257], in the course of carrying out these $\beta$ calculations, also re-evaluated the $P_{XQ}(E, \ell)$ factor of Eq. 4.15 for the FCC probability, using quantum close-coupling calculations and accurate potential curves of Magnier *et al.* [254]. The results showed an FCC probability smaller by a factor of 2 than that estimated in Julienne and Vigué [206] and underscores the necessity of using accurate molecular potential data in collisions calculations. Reference [120] showed that FCC probabilities can be quite sensitive to the molecular potentials.

In a follow-up experiment Marcassa *et al.* [261] used a catalysis laser to study Na trap loss as a function of red detuning. As discussed above, the number of atoms observed in

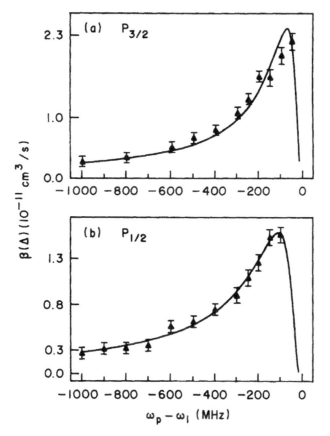

**Fig. 4.16.** Trap-loss spectrum in as Na MOT for catalysis laser detuned to the red from both atomic fine-structure asymptotes, from Marcassa *et al.* [261].

the presence of the catalysis or probe laser is

$$N = \frac{L}{\gamma + \beta(\Delta, I)n_c f}$$

where

$$\beta(\Delta, I) = \beta_t + \beta_c(\Delta)\frac{I}{I_{ref}}.$$

As in Hoffmann *et al.* [177] the intensity of the extra laser is adjusted at each detuning to keep $N$ constant. From this condition Marcassa *et al.* [261] determined $\beta_c(\Delta)$. Figure 4.16 shows the trap-loss spectrum for the catalysis laser detuned to the red of both the $3s\,^2S_{1/2} + 3p\,^2P_{3/2}$ asymptote and the $3s\,^2S_{1/2} + 3p\,^2P_{1/2}$ asymptote. The reason for measuring the trap-loss spectrum from both asymptotes is that both FCC and RE contribute from the $3s\,^2S_{1/2} + 3p\,^2P_{3/2}$ level, while only RE contributes from the lower $3s\,^2S_{1/2} + 3p\,^2P_{1/2}$ level. For comparison, the GP theory is shown in the same figure as a solid line. The overall behavior is in good qualitative agreement with the model. A comparison between the absolute values shows that the amplitude of the trap-loss spectrum to the red of the $3s\,^2S_{1/2} + 3p\,^2P_{1/2}$ asymptote is only 20% lower than the trap-loss spectrum near $3s\,^2S_{1/2} + 3p\,^2P_{1/2}$. Therefore Marcassa et al. [261] concluded that FCC is responsible for only 20% of the trap loss, with RE being responsible for the remaining 80%. This result is in sharp disagreement with the calculations reported in Marcassa et al. [257] in which RE contributes about 30%. These disagreements indicate that the small-detuning trap-loss process and its detailed mechanisms are still poorly understood.

### 4.2.1  Cesium trap loss

Sesko et al. [345] carried out the first collisional trap loss measurements on cesium using a MOT loaded from an atomic beam. In their experiment fluorescence decay was observed after loading. To ensure a constant half-width Gaussian profile for the whole investigated intensity range, the number of trapped atoms was kept low ($\sim 3 \times 10^4$). Figure 4.17 shows the measured $\beta$ as a function of the trap laser intensity obtained by Sesko et al. [345]. Above $4\,\mathrm{mW\,cm^{-2}}$ $\beta$ increases with intensity due to the energy transfer collisions as RE and FCC. Below $4\,\mathrm{mW\,cm^{-2}}$ the dramatic increase in $\beta$ is due to hyperfine-changing collisions (HCC) between ground-state atoms. For Cs, a change from $6s\,^2S_{1/2}(F = 4)$ to $6s\,^2S_{1/2}(F = 3)$ in one of the atoms participating in the collision transfers about $5\,\mathrm{m\,s^{-1}}$ of velocity to each atom. The GP model is also included in Fig. 4.17 (solid line), showing a value a few times smaller than the measurement.

In the same experiment Sesko et al. [345] also investigated the trap loss spectrum using a "catalysis" laser. They did not adjust the catalysis laser intensity at different detunings, as did Hoffmann et al. [177] and Marcassa et al. [259], to maintain constant trap density, but instead measured the trap density at each point on the trap-loss curve, and measured

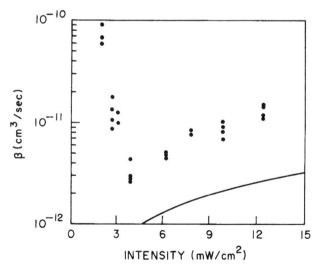

**Fig. 4.17.** Trap loss rate constant $\beta$ as a function of MOT intensity for Cs collisions: a comparison of experiment with GP theory, from Sesko et al. [345].

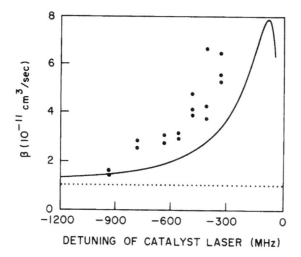

**Fig. 4.18.** Trap loss spectrum in a Cs MOT using a "catalysis" laser (points), and a comparison with GP theory (solid line). The dotted line shows the value of $\beta$ without the "catalysis" laser, from Sesko *et al.* [345].

the $\beta$ decay curve with and without the catalysis laser being present. Their results are presented in Fig. 4.18, together with the GP model (solid line). The scatter in the data reflects the difficulty of obtaining accurate measurement of the trap density. In the late 1990s Fioretti *et al.* [135] directly measured FCC collisional losses from a cesium type I and type II MOT [321] with the trapping light near the $6s\,^2S_{1/2}(F = 4) \rightarrow 6p\,^2P_{3/2}(F = 5)$ and the repumping light near $6s\,^2S_{1/2}(F = 3) \rightarrow 6p\,^2P_{3/2}(F = 4)$ for the type I trap and $6s\,^2S_{1/2}(F = 3) \rightarrow 6p\,^2P_{3/2}(F = 2)$ for the type II trap, by detecting D1 fluorescence emitted on the $6p\,^2P_{1/2} \rightarrow 6s\,^2S_{1/2}$ transition. Measuring the counting rate of D1 photons emitted, the efficiency of detection of both D2 and D1 light, and determining the trap density from the D2 count rate and the spatial extent of the trap, Fioretti *et al.* [135] derive an absolute value for FCC in the cesium MOT. They have examined the sensitivity of FCC to different excited-state hyperfine components by modifying the ground-state hyperfine population that starts the process. They fitted their data with a model that requires molecular hyperfine optical pumping during the collision. Their measurement of $\beta_{FCC}$ ($2 \times 10^{-12}\ \mathrm{cm^3\ s^{-1}}$) and comparison with the previous measurement (Sesko *et al.* [345]) of the *total* $\beta$ ($8 \times 10^{-12}\ \mathrm{cm^3\ s^{-1}}$) leads to the conclusion that FCC contributes about 25% of the total (FCC + RE) loss rate constant. The JV theory predicts about equal contributions from FCC and RE, but it must be borne in mind that the Fioretti *et al.* [135] experiment was carried out under strong-field conditions (total trap intensity as high as $150\ \mathrm{mW\ cm^{-2}}$) and JV is a weak-field theory which neglects hyperfine structure. Therefore it is not too surprising that theory and experiment are not in accord.

### 4.2.2 Rubidium trap loss

Wallace *et al.* [422] carried out the first measurements of trap loss in rubidium, using a MOT similar to that employed by Sesko *et al.* [345] for the study of cesium. The MOT was

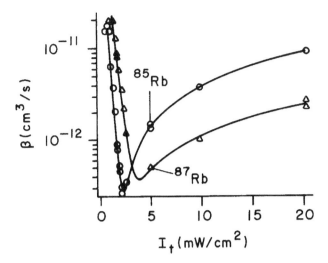

**Fig. 4.19.** Trap loss spectrum for two isotopes of rubidium. The right-hand branch shows trap loss due to FCC and RE. The left-hand branch shows trap loss due to HCC, from Wallace *et al.* [422].

loaded from a rubidium atomic beam slowed with a frequency-chirped diode laser source, and the trap loss determined by monitoring the atomic fluorescence decay. The trap-loss rate constant $\beta$ was obtained by fitting the data to

$$\frac{dN}{dt} = -\gamma N - \beta \int_v n^2 (r, t)\, d^3 r, \tag{4.29}$$

which is just Eq. 4.24 with the loading rate $L$ set to zero. The trap operates at sufficiently low densities ($<2 \times 10^{10}$ cm$^3$ s$^{-1}$) that radiation trapping effects can be ignored [421]. The rate constant for loss due to collisions with a thermal background gas, $\gamma$, was constant and fixed by the background pressure of $\sim 10^{-10}$ torr. The number of atoms in the trap and the density of the trap are determined from: (1) absorption measurements with a weak probe beam; (2) measurement of the excited-state population fraction obtained from selective photoionization of the Rb ($5p\,^2P_{3/2}$) level; and (3) the trap size determined by a digitized CCD camera image. Figure 4.19 shows $\beta$ plotted as a function of total trap intensity at a fixed detuning of 4.9 MHz to the red of the trapping transition for both rubidium isotopes, $^{85}$Rb and $^{87}$Rb. The behavior is similar to that of Cs, Sesko *et al.* [345], with $\beta$ decreasing approximately linearly from high trap intensities to a minimum at a few mW cm$^{-2}$, and then increasing sharply as the trap intensity is further reduced. On the high-intensity side of the curve $\beta$ behavior indicates trap loss due to a combination of RE and FCC, although there is no way to separate the contributions of the two processes in this experiment. At low intensity the sharp increase of $\beta$ is the signature of HCC. In addition to this general behavior, the most striking feature of the high-intensity branch of the trap-loss curve is a strong isotope effect: the value of $\beta$ for $^{85}$Rb is always a factor of 3.3 ($\pm 0.3$) higher than for $^{87}$Rb. Wallace *et al.* [422] point out that this isotope effect was not anticipated by either GP or JV theories; and, since the only significant difference between the two isotopes is their hyperfine structure, it probably results from the influence of long-range molecular hyperfine interactions on the FCC and RE loss mechanisms. The Connecticut group, led by P. L. Gould, also compared their determination of the absolute value of $\beta$ for both isotopes at a trap intensity of 10 mW cm$^{-2}$ to JV theory. The measured values are $\beta = 3.4 \times 10^{-12}$ and $1.0 \times 10^{-12}$ cm$^3$ s$^{-1}$ for $^{85}$Rb and $^{87}$Rb, respectively. The JV theory calculates $\beta = 1.7 \times 10^{-11}$ cm$^3$ s$^{-1}$, factors of 5 and 17, respectively, in disagreement with either experimental measurement. Since Wallace et al. [422] quote an uncertainty of only $\pm 40\%$ in their absolute $\beta$ measurements, the disagreement between experiment and theory is large and real. In addition to ignoring hyperfine structure, it is important to remember that JV is a weak-field theory and that the calculation of the FCC contribution is quite sensitive to $P_{XQ}(E, \ell)$, the accurate calculation of which, in turn, relies on accurate molecular potential energy curves. The strong sensitivity of the FCC probability to the molecular potentials has been examined by Dulieu et al. [120]. Currently, the accord between experiment and theory must be considered unsatisfactory.

On the low-intensity branch of Fig. 4.19, where $\beta$ starts to increase due to HCC, the isotopic ordering of $\beta$ reverses, and the trap intensity at which $\beta$ reaches a minimum differs between the two isotopes by a factor of $\sim 1.5$. Since the ratio of the kinetic energy release due to hyperfine changing collisions from $^{87}$Rb and $^{85}$Rb is

$$\frac{6835 \text{ MHz}}{3036 \text{ MHz}} = 2.25,$$

the factor of 1.5 at first appears surprising. However, Wallace et al. [422] point out that a MOT contains the atoms by position-dependent forces and velocity-dependent forces. If the velocity-dependent forces dominate, and if these forces vary linearly with trap intensity, then one would expect the trap-loss ratio to be equal to the ratio of velocities or

$$\sqrt{\frac{6835 \text{ MHz}}{3036 \text{ MHz}}} = 1.5$$

in accord with observation. At the very lowest trap intensities $\beta$ levels off, and the Connecticut group concludes that under these conditions all the atoms undergoing the HCC process can escape. Therefore the $\beta$ of this plateau is a measure of the HCC collision rate. The measurements show a value of $\sim 2 \times 10^{-11}$ cm$^3$ s$^{-1}$.

At about the same time the Connecticut group was carrying out these rubidium trap loss isotope studies, another group at the University of Wisconsin, led by T. Walker, started investigating $^{85}$Rb. In contrast to Wallace et al. [422], the results reported by Hoffmann et al. [177] were carried out in a vapor-loaded trap, not an atomic-beam loaded trap, and $\beta$ was obtained by fitting Eq. 4.26 to the observed fluorescence loading curves. The Wisconsin group also used a catalysis laser to measure the trap-loss spectrum, i.e. $\beta$ as a function of red detuning of the catalysis laser from the trapping transition. Their results are displayed in Fig. 4.20. They compare their rubidium results to the earlier cesium study (Sesko et al. [345]) and find the following differences: (1) $\beta$ at the peak of the $^{85}$Rb trap-loss spectrum is about a factor of 3.5 smaller than $\beta$ at the peak of the Cs trap-loss spectrum; (2) these peaks do not appear at the same detuning for the two species. Cesium $\beta$ peaks at a detuning of $\simeq 450$ MHz while $^{85}$Rb peaks at $\simeq 160$ MHz. (3) Extrapolating their data to 10 mW cm$^{-2}$

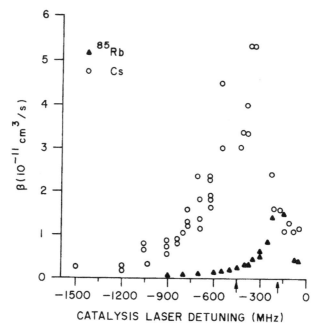

**Fig. 4.20.** Trap loss spectrum of $^{85}$Rb measured by the Wisconsin group and compared to a trap loss spectrum of Cs, from Hoffmann *et al.* [177].

and small detunings, Hoffmann *et al.* [177] determine $\beta = (3.6 \pm 1.5) \times 10^{-12}$ cm$^3$ s$^{-1}$, a factor of 5 below the JV calculation but in quite good agreement with the measurement of Wallace *et al.* [422].

The striking and unexpected isotope effect reported by Wallace *et al.* [422] stimulated further investigation by the Wisconsin group. Feng *et al.* [131] carried out trap-loss studies on both rubidium isotopes similar to those already described by Hoffmann *et al.* [177] for $^{85}$Rb. The trap-loss spectra show that for catalysis laser detunings outside the excited-state hyperfine structure regime, about 400 MHz to the red of the trapping transition, $\beta$ for both isotopes is the same. Within the hyperfine region, Feng *et al.* [131] found about the same ratio of $\beta_{85}/\beta_{87} \simeq 3$ as Wallace *et al.* [422]. Furthermore, they observed a double-peaked structure in the $^{87}$Rb $\beta$ at 400 MHz red detuning and difficulty in making reliable measurements left gaps or holes in the spectrum at detunings near the atomic hyperfine transitions. Figure 4.21 shows these features of the $^{87}$Rb trap loss spectrum. These results clearly demonstrate that the isotope effect as well as the spectral "holes" are associated with trap-loss processes strongly modulated by molecular hyperfine structure. The Wisconsin group detailed their experimental method and summarized the results in Hoffmann *et al.* [176].

The important role of hyperfine structure in trap-loss dynamics demonstrated by the experiments of Wallace *et al.* [422] and Feng *et al.* [131] prompted the development of models proposed by Walker and Pritchard [420] and Lett *et al.* [246]. In essence these

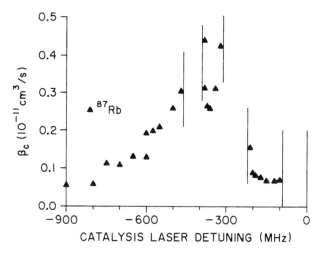

**Fig. 4.21.** Trap loss spectrum of $^{87}$Rb showing "holes" due to hyperfine structure, from Feng *et al.* [131].

studies extend the GP picture of a single attractive excited state to multiple excited states with both attractive and repulsive branches. Figures 4.22(a,b) shows the simplest, illustrative case. Potential curves 0 and $E_1$ represent two closely spaced hyperfine levels of the excited state. The upper curve, 0, shows the familiar $V(R) = -C_3/R^3$ long-range behavior while the lower level splits into $\pm C_3/R^3$ corresponding to an attractive and a repulsive dipole–dipole interaction. Each of the attractive curves gives rise to a GP-like trap-loss spectral profile resulting in an observed "doublet" as a catalysis (probe) laser detunes to the red, indicated in Fig. 4.22(b). The repulsive curve arising from $E_1$ mixes with the attractive curve from 0 in a localized region around the crossing point. As the catalysis laser scans through the crossing, population in the attractive level leaks onto the repulsive curve, effectively rerouting scattering flux away from RE and FS trap-loss processes occurring at smaller $R$. The result is hole burned in the trap loss spectrum shown schematically by the dotted curve in Fig. 4.22(b). Lett *et al.* [246] generalized this model to multiple levels with multiple crossings. In the case of $^{87}$Rb, for example, three hyperfine levels ($F' = 3, 2, 1$) can be excited from the $F = 2$ ground state. Very careful new trap-loss measurements were carried out in $^{85}$Rb right around the detuning region where structure in the trap-loss spectrum was expected. Figure 4.23 shows the results of these new measurements together with the former data of Hoffmann *et al.* [176] and the hyperfine-structure model using $2_u$ potential parameters and the Landau–Zener avoided-crossing probability $P_{LZ} = 0.6$. The agreement between the experimental points and the model appears quite satisfactory. Attempts to fit the data using long-range potentials other than the $2_u$ in the model cannot reproduce the data of Hoffmann *et al.* [177] consistently and these new results reported by Lett *et al.* [246]. Further application of the model to the $^{87}$Rb trap-loss data of Feng *et al.* [131] also shows good agreement, and the model is successful with Cs data (although a printing error in Fig. 8 of Lett *et al.* [246] does not actually permit the comparison in that article; see,

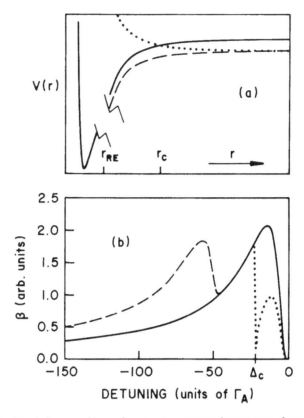

**Fig. 4.22.** Model of the influence of hyperfine structure on trap-loss spectra, from Lett *et al.* [246].

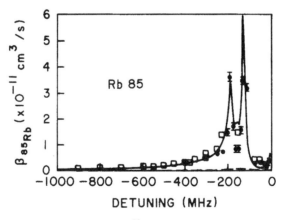

**Fig. 4.23.** Structure in the trap loss spectrum of $^{85}$Rb, showing the influence of hyperfine structure. The solid curve is a model calculation. Full circles are the measurements of Lett *et al.* [246] and the open squares are measurements from Hoffmann *et al.* [177].

however, Corrigendum 1995 *J. Phys. B: At. Mol. Opt. Phys.* **28**, 1). The application of this hyperfine-level avoided-crossing model leads to two major conclusions: (1) structure in the trap-loss spectrum is due to excitations of multiple hyperfine levels and avoided crossings between them and (2) the major trap loss mechanism is long-range RE on the $2_u$ potential curve. However, we must caution that using the molecular Hund's case (c) symmetry label $2_u$ for potential curves in a region of strong hypefine-induced mixing is unwarranted. These results do suggest that a state characterized by weak optical coupling is implicated in the trap-loss mechanism.

Peters *et al.* [311] greatly simplified the interpretation of trap-loss data by measuring $\beta$ from collision partners excited near the $Rb(5p\,^2P_{1/2}) + Rb(5s\,^2S_{1/2})$ asymptote. Since most hyperfine-level curve crossings are eliminated and the spacing between the levels is large, all of the attractive hyperfine levels converging on the $Rb(5s\,^2S_{1/2}; F = 2) + Rb(5p\,^2P_{1/2}; F' = 1)$ will closely follow a $C_3/R^3$ power law and their contributions to the total trap loss rate constant should be simply additive. Furthermore, trap loss contributions to the red of the $5p\,^2P_{1/2}$ asymptotes can only be due to the radiative escape process since energy conservation closes the FCC channel. The results shown in Fig. 4.24 confirm that under these simplified conditions there is no discernible isotope effect between $^{85}Rb$ and $^{87}Rb$, that the GP model works fairly well, but that multiple orbiting of the collision partners must be included to obtain good agreement between the measurements and the model. Another benchmark is the position of the peak in the trap-loss spectrum. For hyperfine levels

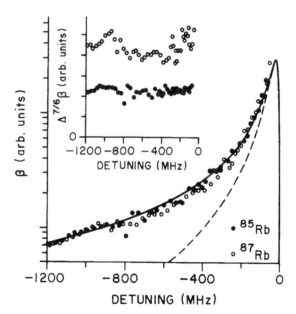

**Fig. 4.24.** Rubidium trap loss spectra from the Wisconsin group showing that no isotope effect is measurable for trap loss collisions from the $5p^2\,P_{1/2} + 5s^2\,S_{1/2}$ asymptote and that multiple orbiting must be included in any model of the trap loss process. The solid line is a model calculation including multiple orbits and the dashed line shows the same model with a single orbit, from Peters *et al.* [311].

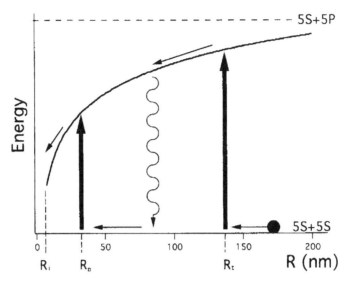

**Fig. 4.25.** Schematic of flux enhancement. The first excitation at $R_t$ puts flux on the long-range $C_3/R^3$ attractive potential. This excitation effectively pulls added flux into the small-$R$ region where it is probed by the second excitation at $R_p$, from Sanchez-Villicana *et al.* [336].

converging on the Rb($5p\,^2P_{3/2}$) + Rb($5s\,^2S_{1/2}$) asymptote the GP model predicts this peak to be near a red detuning of about 115 MHz. Unfortunately this is also the region where curve-crossing and interactions among the hyperfine levels complicate the trap-loss spectrum. From the calculated shapes of the long-range potentials converging on the Rb($5s\,^2S_{1/2}$; $F = 2$) + Rb($5p\,^2P_{1/2}$; $F' = 1$) asymptote, however, Peters *et al.* [311] calculate that the trap-loss peak should occur at about 25 MHz red detuning with no complicating hyperfine interactions to cloud the model prediction. Probing the trap with the "catalysis" laser so close to the trapping transition unfortunately leads to such strong trap perturbations that trap loss measurements become unreliable. Therefore the Wisconsin group could not apply their measurement technique to test this particular prediction of the GP model.

Sanchez-Villicana *et al.* [336] have demonstrated a flux enhancement effect in ultracold collisions by comparing trap loss rate constants in an atomic-beam-loaded rubidium MOT with trap and probe lasers present simultaneously or alternately and with the probe laser tuned as far as 1 GHz to the red of the MOT trapping transition. Figure 4.25 outlines the basic process. Scattering flux enters the collision on the 5S + 5S ground state and is excited by the MOT laser at $R_t$ (red detuning $\sim\Gamma$) to one of the attractive $-C_3/R^3$ potentials converging on the 5S + 5P asymptote. The excited flux accelerates and spontaneously relaxes back to the ground state with increased kinetic energy, enabling higher partial wave components to overcome their angular momentum barriers at closer internuclear separation. At some near-zone Condon point $R_C$ (red detuning $\sim$50–150 $\Gamma$) a probe laser re-excites the flux to the $-C_3/R^3$ potential where it may undergo an inelastic trap loss interaction (e.g. a short-range curve crossing leading to a fine-structure change). Sanchez-Villicana *et al.* [336]) measure an enhancement in the trap loss constant $\beta$ due to the combined effect of the trap and probe lasers by defining an enhancement factor $\eta$,

$$\eta = \frac{\beta_{t+p} - \beta_t}{\beta_p},\qquad(4.30)$$

which is the difference between the trap loss constant with both beams present simultaneously or alternately. Note that if there is no combined effect, then $\eta$ would remain equal to one. This flux enhancement process is significant because it might influence the results of any experiment where trap and probe beams are present simultaneously. The Connecticut group led by P. Gould has also reported a comprehensive study of trap loss collisions for $^{85}$Rb and $^{87}$Rb covering a wide range of trap parameters. The relevant reference is Gensemer, *et al.* [150].

In a pair of articles [255, 263] Marcassa and coworkers have investigated the relative contribution of fine-structure-changing collisions to total trap loss in $^{85}$Rb collisions. The basic idea is to measure the 5p $^2$P$_{1/2}$ population arising from FCC collisions between 5p $^2$P$_{3/2}$ + 5s $^2$S$_{1/2}$ atoms in a typical Rb MOT. The population is detected by first exciting a bound–bound transition in the atom, 5p $^2$P$_{1/2}$ + $\hbar\omega \rightarrow$ 8s $^2$S$_{1/2}$, followed by selective photoionization out of the 8s level. In the first of the two-part series, the authors reported a contribution from FCC collisions to the total trap loss rate constant of only about 4%; but a quantitative determination was hampered by the fact that a narrow-band continuous laser ($\Delta\omega \simeq 2\pi \times 1$ MHz), was used to excite the 5p $^2$P$_{1/2}$ population. Since the Doppler width was considerably greater ($\simeq 600$ MHz), the entire population could not be excited simultaneously. In a follow-up report, the authors substituted a pulsed dye laser with a spectral width covering the entire Doppler width of the 5p $^2$P$_{1/2}$ + $\hbar\omega \rightarrow$ 8s $^2$S$_{1/2}$ transition. The count rate per pulse was greatly improved over the continuous-wave (cw) laser excitation, and the authors reported a revised fractional contribution of the FCC process to the total trap loss rate of $\simeq 35$%. A fully satisfactory theory of trap loss, predicting (or at least rationalizing) the relative contributions of RE and FCC contributions has not yet been developed.

### 4.2.3 Lithium trap loss

Although collisional trap loss occurs for all alkali species, lithium shows two unique features. First, the excited-state hyperfine structure is inverted with 2p $^2$P$_{3/2}(F' = 3)$ lying lower than 2p $^2$P$_{3/2}(F' = 2, 1, 0)$. Therefore long-range attractive molecular states correlating to the 2s $^2$S$_{1/2}(F = 2) +$ 2p $^2$P$_{3/2}(F' = 3)$ asymptote are less likely to be perturbed by hyperfine interactions and avoided crossings with higher-lying molecular states. Secondly, the fine-structure splitting, $\Delta E_{FS}$, between the 2p $^2$P$_{3/2}$ and 2p $^2$P$_{1/2}$ levels is sufficiently small that, by varying the trapping laser intensity, the MOT trap depth can be made comparable to the FCC energy release. In temperature units $\Delta E_{FS}/k_B$ is only 0.48 K. A fine-structure-changing collision results in each atom leaving the binary encounter with one-half the exoergicity. If the trap depth $E_T$ is shallower than this $\frac{1}{2}\Delta E_{FS}$, then each atom can escape the trap and FCC becomes an important loss mechanism. However, if the trap deepens below $\frac{1}{2}\Delta E_{FS}$, the atoms cannot escape and the FCC mechanism no longer functions as a

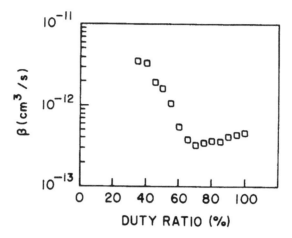

**Fig. 4.26.** Trap loss rate constant $\beta$ as function of "trap depth" (MOT duty ratio) for lithium atoms. The rapidly rising left-hand branch is thought to be due to FCC and RE, but as the trap becomes deeper (increasing duty ratio) FCC no longer contributes, from Kawanaka *et al.* [213].

trap loss process. In this situation, the remaining loss from radiative escape may be studied without the influence of FCC.

Kawanaka *et al.* [213] reported the first measurement of $\beta$ for $^7$Li. The experiment implemented an unusual four-beam MOT in which three of the beams point toward the vertex of an equilateral pyramid and the fourth beam counterpropagates along the axis of the pyramid, itself aligned with a trap-loading Li beam. The axial counterpropagating laser beam serves the double role of atomic beam slower and MOT confining laser. The total intensity in the MOT laser was 30 mW cm$^{-2}$, and to change the trap depth, the MOT beams were on–off modulated at 500 kHz with the "on" time varying from $\sim$30 to 100 per cent duty ratio. Values of $\beta$ as a function of MOT duty ratio were extracted from curve fitting fluorescence decay. The trap was operated with a small number of atoms, such that the spatial density profile was always close to a Gaussian profile. As the duty ratio (or trap depth) was varied, $\beta$ exhibited the behavior shown in Fig. 4.26. The value of $\beta$ appears to show a plateau $\sim$3 $\times$ 10$^{-12}$ cm$^3$ s$^{-1}$ at the minimum duty ratio of $\sim$30 per cent (shallow trap). As the duty ratio increases (deeper trap) this value decreases rapidly by about one order of magnitude, reaching a minimum value of $\sim$3.2 $\times$ 10$^{-13}$ cm$^3$ s$^{-1}$ at $\sim$65 per cent duty ratio. As the MOT laser "on" duty ratio increases above this minimum point to 100 per cent, $\beta$ increases only slightly to $\sim$5 $\times$ 10$^{-13}$ cm$^3$ s$^{-1}$. Kawanaka *et al.* [213] interpret this $\beta$ behavior as showing the contributions of both FCC and RE below the minimum, but at higher duty ratios (deeper traps) the FCC process no longer contributes. Therefore above about 65 per cent duty ratio the only source of collisional trap loss is radiative escape. If this interpretation is correct, then the duty ratio at which the minimum $\beta$ occurs must be equivalent to $E_T = \frac{1}{2}\Delta E_{FS}$. Kawanaka *et al.* [213] carried out a direct measurement of the trap depth at 100 per cent duty ratio by applying the kick-and-recapture technique described by Raab *et al.* [321]. This measurement showed that the MOT is anisotropic with a depth of 0.64 K along the radial direction and 1.3 K along the axial direction. The trap-modulation technique effectively varies the average MOT intensity and therefore the average trap depth. A MOT operates with two kinds of restoring forces: proportional to position and proportional to velocity. If the position-dependent conservative force dominates, then the trap depth should be linear with the average MOT intensity. If the velocity-dependent dissipative force dominates then the trap capture ability should vary as the square root of the average MOT intensity. Given the direct measurement of the trap depth at 100 per cent duty ratio and assuming a square-root dependence, $\frac{1}{2}\Delta E_{FS}$ corresponds to about 60 per cent duty ratio, near the minimum $\beta$ and roughly consistent with the interpretation of an RE-only trap loss at duty ratios above 60–65%. It should be borne in mind that the anisotropy of the trap was not explicitly taken into account in these measurements. It should be further noted that value of $\beta$ reported in the RE-only regime, $\sim 5 \times 10^{-13}$ cm$^3$ s$^{-1}$, is about one order of magnitude greater than predicted either by the GP or JV models.

Ritchie *et al.* [329] carried out another study of collisional trap loss in a $^7$Li MOT, measuring $\beta$ over a wider range of intensities and several detunings. This study carefully took account of the anisotropy of the MOT trap depth by modelling and simulating atom escape velocities as a function of angle, intensity, and detuning. In a separate study Ritchie *et al.* [328] developed and examined this model in detail. The MOT set-up used the conventional six-way orthogonal laser beam arrangement and was loaded from an optically slowed atomic beam. The trap loss rate constant $\beta$ was extracted from fits to fluorescence decay after the loading was shut off. Figure 4.27 shows the measured $\beta$ for four values of laser detuning and a range of intensities. For each of the four detunings the measured loss rate was largest at small intensities and decreased with MOT intensity until reaching a distinct minimum, which occurred at a different intensity for each detuning. Similar to the interpretation of Kawanaka *et al.* [213], this $\beta$ minimum is attributed to the suppression of FCC loss. At higher values of intensity the loss is due entirely to RE. Also, shown in Fig. 4.27 are the results of an OBE-theory calculation [207] for RE trap loss rates in Li. The OBE approach is appropriate because the MOT intensities in this study reach as high as 120 mW cm$^{-2}$ where weak-field theories such as GP or JV cannot be applied. In addition the OBE theory was found not to fail for Li as it does for the heavier alkalis (Julienne *et al.* [207]). The OBE theory predicts a strong dependence on trap depth with $\beta_{RE}$ scaling as $E_T^{-3.0}$, and therefore a thorough understanding of trap depth spatial and intensity dependence is crucial. To that end Ritchie *et al.* [328] developed a model and carried out trajectory simulations of Li atoms subject to various conditions of intensity and detuning in the MOT. These simulations revealed that the trap depth $E_T$ was highly dependent on $\theta$, $\phi$ the polar and azimuthal angles with respect to the MOT symmetry axis. From the model spatial dependence Richie *et al.* [329] were able to calculate $\beta$ averaged over the different trap directions,

$$\beta = \frac{\beta_0}{4\pi} \int \left(\frac{E_0}{E_T}\right) d\Omega \tag{4.31}$$

where $\beta_0$ and $E_0$ are the trap loss rate constant and trap depth in the shallowest direction, respectively. The result showed a rather startling anisotropy with $\beta/\beta_0 = 0.23 \pm 0.02$, relatively independent of intensity and detuning. The graphs in Fig. 4.27 plot this averaged $\beta$ as well as the calculated intensity $I_c$ required to recapture an atom released in the shallowest

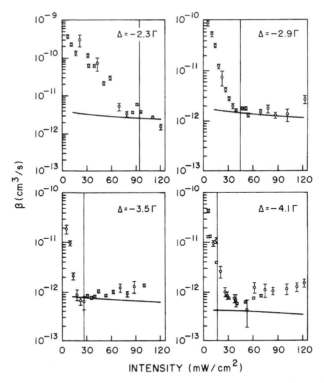

**Fig. 4.27.** Trap loss rate constant $\beta$ for four different detunings and a range of intensities in the Rice group's Li MOT. The vertical lines in each plot denote $I_c$ the calculated critical intensity required to recapture an atom released in the shallowest direction, from Ritchie *et al.* [329].

direction with a kinetic energy equal to $\frac{1}{2}\Delta E_{FS}$. Except for the largest detuning, the agreement between the calculated $I_c$ and the breakpoint in the $\beta$ vs. $I$ plots (where $\beta_{FCC} \to 0$) inspires confidence in the basic correctness of the model of Ritchie *et al.* [328]. In contrast, while the OBE calculation gives approximately the correct value for $\beta_{RE}$ near $I_c$, the predicted and observed trap depth dependences go in opposite directions. The culprit is probably molecular hyperfine structure, which is ignored in the application of the OBE theory. At the time of writing the disagreement remains unresolved. It is, however, interesting to note that Ritchie *et al.* [328] find that the increase in $\beta_{RE}$ with MOT intensity above the breakpoint can be well fitted to a function proportional to the product of ground- and excited-state atomic fractions and that this fitting works over the range of detunings from $\Delta = -2.3\Gamma$ to $-4.1\Gamma$. Such a fitting is more consistent with a GP or JV picture of radiative escape since the quasistatic theories should not be very sensitive to trap depth or detuning. At the high trap intensities of these experiments, however, the basic weak-field assumption of GP and JV cannot possibly be appropriate and the "agreement" is probably just fortuitous. Further work in the application of the theory is clearly needed.

### 4.2.4 Potassium trap loss

Potassium is the last stable atom in Group IA of the periodic table to be successfully confined in a MOT and on which cold collisions have been investigated. The peculiarities of K atomic structure have precluded a conventional approach to trap-loss measurements. Williamson and Walker[448] reported the first successful potassium MOT and the first studies of exoergic trap loss. The Wisconsin group trapped two isotopes, $^{39}$K and $^{41}$K, with natural isotopic abundances of 93 and 7 per cent, respectively; but the very small excited-state hyperfine splittings of these isotopes required a modification of conventional MOT technique. The unusual fact is that, in the case of $^{39}$K, the splitting of the entire excited-state hyperfine manifold is only 33 MHz, and the separation between $4p\,^2P_{3/2}(F' = 3)$ and $4p\,^2P_{3/2}(F' = 2)$ is only 21 MHz. In the case of $^{41}$K the excited hyperfine manifold covers only 17 MHz. With a natural line width of 6.2 MHz, overlapping transitions from the $4s\,^2S_{1/2}(F = 2)$ ground state to adjacent hyperfine levels would lead to considerable optical pumping and heating as spontaneous emission populated the lower $F = 1$ ground state. To overcome this problem Williamson and Walker [448] tuned the trapping and repumper lasers to the red of the entire hyperfine manifold, essentially treating the $^2P_{3/2}$ fine-structure level of $^{39}$K as an unresolved line with $\sim$30 MHz width. Although optical pumping "population leakage" will certainly be higher when the maximum $F|M_F|$ state cannot be populated uniquely, this tactic has at least the advantage that the capture range of K atom velocity extends to 30 m s$^{-1}$, an increase of about a factor of 5 over the capture range for Na. The increase in velocity capture range increases the MOT loading rate and partially compensates for optical pumping losses. The Wisconsin potassium MOT operates at high light intensity (470 mW cm$^{-2}$ for $^{39}$K) and with a large number of trapped atoms such that the loading proceeds mostly at constant density. For $^{41}$K the MOT intensity goes as high as 530 mW cm$^{-2}$ with a large beam diameter of 0.6 cm. Fitting trap-loading fluorescence curves to Eq. 4.26 yields the total loss rate constant $\Gamma(n) = \gamma + \beta n$, where $\gamma$ is the rate constant for collisions with a hot background gas and $\beta$ is the familiar trap loss rate constant for ultracold potassium collisions. The $\beta$ rate constant is then determined by measuring the *density dependence* of the total trap loss rate, $\Gamma$ at fixed detuning. The slope of this plot, d$\Gamma$/d$n$ yields $\beta$, and this program is carried out over a range of detunings. Figure 4.28 shows the variation of $\beta$ with the detuning $\Delta$ for the $^{39}$K MOT. Although some variation of $\beta$ with detuning is evident, the range is considerably narrower than the typical trap-loss spectra reported earlier by the Connecticut and Wisconsin groups for rubidium. Obtaining the detuning dependence of $\beta$ from the density plots is clearly much more arduous than the using a catalysis laser, so the difference in detuning range is not surprising. Williamson and Walker [448] report that $^{41}$K shows even less detuning dependence than $^{39}$K. The trap loss collision dynamics of K is much less well-studied than the other alkalis and investigations must continue to clarify the role played by the differences in hyperfine structure.

Another study of the dependence of the trap loss rate of a potassium MOT as a function of trap laser intensity was recently performed by Santos *et al.* [340] using a MOT loaded from vapor. The loading measurements were carried out under conditions of constant trap density. To be able to measure a large range of intensities without compromising the trap

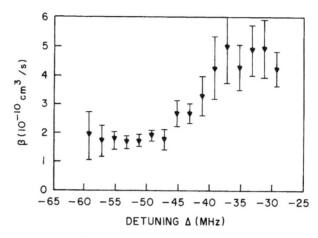

**Fig. 4.28.** Trap loss spectrum of $^{39}$K in the potassium MOT of the Wisconsin group, from Williamson and Walker [448].

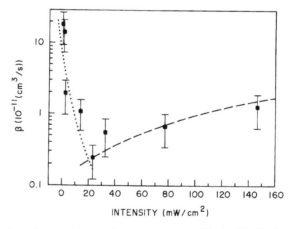

**Fig. 4.29.** Intensity dependence of the trap loss rate constant $\beta$ in the São Carlos postassium MOT, from Santos *et al.* [340].

performance, the sudden change of intensity technique was used: the trap was loaded at high intensity, after which a neutral density filter was mechanically introduced, producing a low-intensity condition. The transition between the two intensities regimes causes a transient decay in the number of atoms that permits a determination of $\beta$. The values of $\beta$ measured for several intensities are plotted in Fig. 4.29, where each point is a result of 15 independent measurements. From high intensities down to $10\,\mathrm{mW\,cm^{-2}}$, $\beta$ shows only a small variation with a possible tendency of increasing with intensity. In this regime, indicated by a dashed line in Fig. 4.29, RE and FCC are dominant losses. At about $10\,\mathrm{mW\,cm^{-2}}$ an abrupt increase in $\beta$ is observed, which is associated with HCC as observed in other systems. The HCC regime for trap loss is indicated in Fig. 4.29 as a dotted line.

### 4.2.5 Sodium–potassium mixed-species trap loss

All experiments and theories concerning exoergic collisions in atomic traps have considered only homonuclear systems. Santos *et al.* [339] recently demonstrated that two different alkalis (Na/K) could be cooled and confined in the same MOT. This experiment opens the study of heteronuclear ultracold exoergic collisions.

The mixed-alkali MOT of Santos *et al.* [339] uses a glass vapor cell that contains a mixture of Na/K vapor at room temperature. The trapping laser beams for both alkalis (589 nm for Na and 766 nm for K) are combined using dichroic mirrors and enter the MOT cell copropagating. The optimum trapping conditions are similar to each alkali individually considered. Fluorescence detection measures the number of trapped atoms and a CCD camera measures the spatial distribution of each species. The sodium atomic cloud is located within the larger potassium cloud. To investigate the inter-species cold collision effects, Santos *et al.* [339] measured the loading of the sodium trap in the presence and absence of the cold potassium cloud. As the sodium trap loaded, the net loading rate is expressed as,

$$\frac{dN_{Na}}{dt} = L - \gamma N_{Na} - \beta n_{Na} N_{Na} - \beta' n_K N_{Na}, \tag{4.32}$$

where $\gamma N_{Na}$ is the collision rate of trapped Na with the hot background gas (both Na and K). Observing the transient behavior in a regime of constant density for both species, one can measure $\beta'$, which is the cross-species trap-loss collision rate. The measured value for Na/K was $\beta' = 3.0 \pm 1.5 \times 10^{-12}$ cm$^3$ s$^{-1}$. The loss rate constant for Na/Na collisions was previously reported by Marcassa *et al.* [257] to be $\beta = 3 \pm 1 \times 10^{-11}$ cm$^3$ s$^{-1}$, about one order of magnitude larger than $\beta'$ due to Na/K collisions. The Brazilian group, led by V. S. Bagnato, rationalized the marked difference by noting that the long-range, excited-state molecular potential in NaK arises from a *nonresonant* dipole–dipole interaction and varies only as $-C_6/R^6$ instead of the familiar $-C_3/R^3$ as in the homonuclear case. They argued that the Na/K atom pairs excited at long range will experience much less acceleration along their line of centers with a consequent reduction in the survival probability factor (see, for example, Eqs. 4.14 or 4.17) of the rate constant. An alternative interpretation could be that the $-C_6/R^6$ interaction implies a much reduced phase space available for the initial excitation due to the much smaller Condon point.

### 4.2.6 Sodium–rubidium mixed-species trap loss

In subsequent experiments the São Carlos group headed by L. Marcassa investigated a series of mixed species traps, beginning with the sodium–rubidium combination [382]. Superposing the Na and Rb MOTs, they found the Na MOT volume to be smaller than the Rb MOT volume by about a factor of 8, although the densities were the same to within a factor of 2 ($\simeq 10^{10}$ cm$^{-3}$). They carried out loss-rate studies using the basic rate equation,

$$\frac{dN_{Na}}{dt} = L - \gamma N_{Na} - \beta \int_v n_{Na}^2 \, d^3r - \beta' \int_v n_{Na} n_{Rb} \, d^3r \tag{4.33}$$

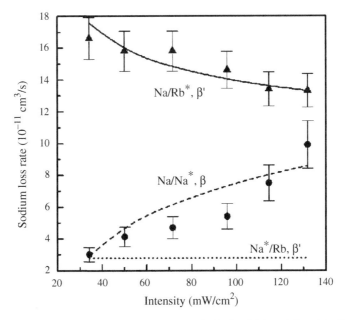

**Fig. 4.30.** Plot of collisional trap loss rate constants for Na–Na* collisions ($\beta$) and for Na–Rb* collisions ($\beta'$) as a function of Na MOT intensity.

in which $\beta$ is the loss rate constant for the Na MOT in the absence of Rb and $\beta'$ loss rate with the Rb MOT present. Since the Na MOT volume is much smaller than the Rb MOT, and since the Na MOT density obeys a Gaussian spatial distribution, $n(r, t) = n_0(t)e^{-2(r/w)^2}$ the rate expression can be written as

$$\frac{dN_{Na}}{dt} = L - \gamma N_{Na} - \frac{2\beta N_{Na}^2}{w^3\pi^{3/2}} - \beta' n_{Rb} N_{Na}. \tag{4.34}$$

The two rate constants can be determined separately, first by blocking the Rb MOT and measuring the loss rate due only to Na collisions; then, with $\beta$ determined, measuring the loss rate again and using Eq. 4.34 to fit $\beta'$. Figure 4.30 shows the results of the measurement of $\beta$ and $\beta'$ as a function of Na MOT intensity. At first it may appear surprising that $\beta'$ decreases with increasing intensity while $\beta$ increases. Telles *et al.* explain this behavior by pointing out that, as the intensity in the Na MOT decreases, the barrier height of the Na trap decreases whereas the Rb excited-state density changes very little. Collisions between excited Rb and ground-state Na leading to radiative escape will therefore occur with increasing probability. The detuning dependence of the two loss rate constants was also studied and found to differ markedly from the prediction of a simple Gallagher–Pritchard (G-P) model calculation. Figure 4.31 shows the measurement of the detuning dependence of loss rate constants together with the predictions of the G-P model. These measurements are carried out by detuning a "catalysis laser" while keeping the number of Na and Rb atoms constant. The results, shown in Fig. 4.31 do not agree at all with the conventional

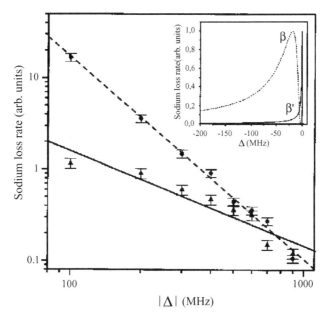

**Fig. 4.31.** Detuning dependence of $\beta$ and $\beta'$. Note that $\beta'$ exhibit a slope of $-2$ on the log–log plot, indicating a measured inverse-square dependence of $\beta$ on detuning.

prediction which indicates a weak detuning dependence. Figure 4.31 shows a decreasing $\beta'$ which falls off as the square of the detuning. In a follow-up report [379] Telles *et al.* resolve the discrepancy by invoking a new loss channel, collisions between excited Na and excited Rb atoms.

### 4.2.7   Other mixed-alkali loss measurements

#### *Na–Cs trap loss*

Following up on an earlier report of photoassociative ionization in the NaCs heteronuclear system [347], Shaffer *et al.* [348] have carried out a study of trap loss when a Cs MOT and an Na MOT are superposed. They showed by varying the MOT light intensity in the Na and Cs MOTs independently and examining the loading rates of each trap with and without the presence of the other constituent, that trap loss processes due to the Na–Cs interaction were dominated by the fine-structure change in excited Na. Adapting the analysis of Dasheveskaya *et al.* [101], discussed earlier in Section 4.1, to the heteronuclear case, they ascribed this release to either Coriolis coupling of the $2\,^3\Pi_{1,2}$ Na–Cs levels with the $2\,^3\Pi_{0\pm}$ levels or spin–orbit mixing of the $2\,^3\Pi_1$ and $2\,^1\Pi_1$ levels as indicated in Fig. 4.32. Measurement of the trap loss rate constants were insensitive to the density of excited Cs, from which Shaffer *et al.* [348] conclude that Cs fine-structure-changing collisions do not contribute significantly to trap loss in the mixed MOT. They argue that spin–orbit and Coriolis coupling energies are too small to compete with the strong fine-structure splitting

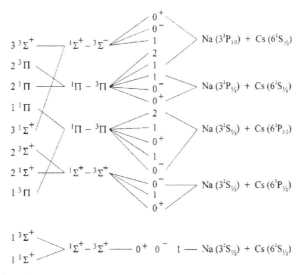

**Fig. 4.32.** Correlation diagram of the molecular states constructed from the ground and first excited fine-structure states of Na and Cs.

in atomic Cs, and therefore collisional interaction with the Na atom does not produce the energy release associated with the $Cs(^2P_{3/2} \rightarrow {}^2P_{1/2})$ transition.

### K–Rb trap loss

In a continuation of their systematic study, the São Carlos group has measured trap loss in two superposed MOTs of Rb and K. This combination is especially interesting because the energy asymptotes for the combinations

$$Rb^*(5p) + K(4s)$$

and

$$K^*(4p) + Rb(5s)$$

are separated by only about $250\ cm^{-1}$, the most near-degenerate pair of all the heteronuclear alkali combinations. Therefore the long-range van der Waals interaction should be comparatively strong, rendering those molecular states associated with $Rb^*(5p) + K(4s)$ long-range attractive and those associated with $K^*(4p) + Rb(5s)$ long-range repulsive. One might expect therefore that this heteronuclear combination would optimize the formation rate, by photoassociation, of translationally cold, bound RbK molecules. Although the search for cold RbK still continues, the group has reported trap loss measurements in which they demonstrated that excited Rb atoms resulted in significant mixed trap loss. Marcassa *et al.* [262] considered radiative escape rather than fine-structure change to be the principal loss mechanism, notwithstanding the fact that collisional interaction, even in this most favorable case, is much shorter range than the $C_3/R^3$ resonant dipole interaction of a homonuclear

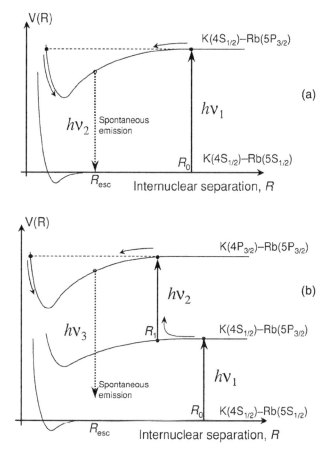

**Fig. 4.33.** Schematic potential curves showing attractive entrance channels for singly excited and doubly excited collision partners in KRb.

pair; and consequently the time for spontaneous emission to take place during the course of the collision is much reduced. However, in contrast to the Na–Cs case studied by Shaffer *et al.* [348], where fine-structure change in excited Na was shown to control the heteronuclear trap loss, fine-structure trap loss in the collision entrance channels associated with Fig. 4.33(a) is unlikely since the Rb fine-structure splitting is quite large ($237 \text{ cm}^{-1}$) and therefore not susceptible to significant mixing and recoupling during a collision with K in the ground state. The São Carlos group suggest that collisions in the doubly excited entrance channel (Fig. 4.33b) may be responsible for the mixed trap loss. Their argument relies primarily on the relative size of the trap loss rate constant calculated from a Gallagher–Pritchard model assuming a $C_6/R^6$ interaction potential, corresponding to the entrance channel of Fig. 4.33(a), or a $C_5/R^5$ potential corresponding to the doubly excited entrance channel of Fig. 4.33(b). The longer range of the $C_5/R^5$ potential results in a greater calculated mixed trap loss rate constant, assuming of course radiative escape to be the dominant mechanism. Although the measurements definitely show that trap loss from a heteronuclear interaction

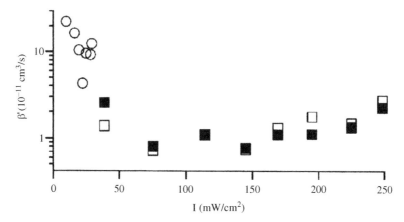

**Fig. 4.34.** Composite plot of the trap loss rate constants measured by the Pisa and São Carlos groups [381]. The circles are the data from the Pisa group and the squares (filled and open) correspond to two different data sets from the São Carlos measurements.

depends on the presence of excited Rb, the corresponding trap-load studies, with the roles of the K and Rb traps reversed, was not carried out; and therefore the sensitivity to excited K of the heteronuclear trap loss cannot be unequivocally determined. Without this piece of the puzzle it is difficult to gauge the relative importance of the singly or doubly excited entrance channels of Fig. 4.33.

### Rb–Cs trap loss

In a joint study between the São Carlos group and the Pisa group headed by E. Arimondo, trap loss studies in a mixed Rb–Cs trap have been carried out [381]. The two groups measured the time dependence of loading curves for Rb with and without the presence of cold Cs atoms. The Pisa group studied $^{87}$Rb at relatively low trapping powers (10–45 mW cm$^2$) while the São Carlos group used $^{85}$Rb over a wider intensity range (40–250 mW cm$^2$). A composite plot of the results for the trap loss rate constant over the range covered by the two experiments is shown in Fig. 4.34. The low-intensity rise is probably due to hyperfine changing collisions while at higher intensities the trap loss rate constant varies very little. The absolute values of the rate constants are smaller by about a factor of 5 than those measured in the K–Rb study.

### Li–Cs trap loss

At Heidelberg the group led by M. Weidemüller has studied trap loss from heteronuclear collisions of Li and Cs in a superposed trap [342]. Since these are the two alkalis at opposite ends of the group IA column of the periodic table, they have widely separated D line resonance excitation energies: 1.85 eV for Li and and 1.45 eV for Cs. One therefore expects that collisions entering from the Li* + Cs asymptote will encounter repulsive potentials while the Li + Cs* asymptote will follow attractive curves. Therefore Li* should play no

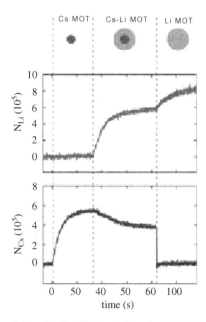

**Fig. 4.35.** Sequential loading of first the Cs MOT and then the Li MOT. The middle panel shows the decrease in Cs MOT density with increasing Li loading. The right-hand pane shows Li reaching maximum density after the Cs MOT is cut off. The Cs MOT is much smaller spatially than the Li MOT because of the much lower average velocity in the Cs MOT.

role in trap loss collisions, and that is indeed what Schlöder *et al.* [342] find. Figure 4.35 shows the time dependence of loading first a Cs MOT, the subsequent decrease in Cs MOT density due to a superposed Li MOT, and finally the increase in Li MOT density due to the cut off of the Cs MOT. By varying independently the intensity and detuning of the two MOTs the Heidelberg group has shown that the heteronuclear loss is due only to Cs* collisions, although the actual mechanism of the loss, radiative escape or fine-structure change was not distinguished.

### 4.2.8 Rare-gas metastable loss in MOTs and optical lattices

Rare gas metastable atoms can also be trapped in MOTs and give rise to trap loss by Penning or associative ionization collisions. Bardou *et al.* [36] have carried out an unusual variation on the trap loss experiment by decelerating a beam of helium metastable atoms, $He(2\,^3S_1)$, and capturing the slowed atoms in a MOT. The cooling transition is between the $2\,^3S_1$ "ground state" and the $2\,^3P_2$ excited state ($\lambda = 1.083\,\mu m$). They carried out trap-decay measurements by shutting off the loading beam and observing the non-exponential decay of the *ions* rather than the usual fluorescence. By fitting the usual trap-loss equations to the ion decay curves, they determined a trap loss rate from Penning ionization,

$$He^* + He^* \rightarrow He + He^+ + e. \tag{4.35}$$

The MOT contains a significant fraction of $2\,^3P_2$ as well as $2\,^3S_1$ so the He* collisions could be between any combination of these two species. Bardou *et al.* [36] determine $\beta = 7^{+21}_{-5} \times 10^{-8}$ cm$^3$ s$^{-1}$ which is a remarkably big number. In fact, the formula (Eq. 48 given in Julienne and Mies [202]), calculated from the $S$-matrix unitarity condition, shows that an upper bound to the Penning ionization rate constant for collisions between He $^3S_1$ atoms approaching on the $^1\sum^+_g$ molecular state will be $2 \times 10^{-9}$ cm$^3$ s$^{-1}$ at $T_{MOT} = 1$ mK. The fact that the measured $\beta$ is more than an order of magnitude higher than the s-wave unitarity limit may mean that collisions between He $^3S_1$ and He $^3P_2$ play an important role. Julienne *et al.* [204] confirmed this conjecture by showing that the unitarity limit increases by a factor of $(\ell_{max} + 1)^2$, where $\ell_{max}$ partial waves are captured by the excited state. At 1 mK $\ell_{max}$ is about 5. Bardou *et al.* [36] probed the role of the excited state by modulating the trapping light and observing the effect on the ion production rate. The decline was dramatic, and by varying the density in the MOT they were able to determine $\beta_P = 14 \times 10^{-8}$ cm$^3$ s$^{-1}$. Therefore their measured $\beta = 7^{+21}_{-5} \times 10^{-8}$ cm$^3$ s$^{-1}$ is an average of $\beta_S$ and $\beta_P$, weighted by the product of the relative ground-state population for S–S collisions and the product of the ground- and the excited-state population for S–P collisions. Both kinds of collisions contribute a loss rate constant several orders of magnitude greater than FCC or RE in the alkalis. As a consequence [36] report a steady-state density in the MOT of about $10^8$ cm$^{-3}$, about two orders of magnitude lower than the density of a typical alkali MOT. Mastwijk *et al.* [264] have reported a new measurement of the rate constant for Penning and associative ionization in metastable He trapped in a MOT. Their reported result, $(1.9 \pm 0.8) \times 10^{-9}$ cm$^3$ s$^{-1}$, is about two orders of magnitude smaller than that of [36]. At the time of writing, to the best knowledge of the author, this discrepancy has not been resolved. Katori and Shimizu [212] and Walhout *et al.* [418] have also measured trap-loss rates of ionizing collisions in metastable Kr and Xe traps.

An interesting variation on "trap loss" in ionizing collisions is the modification of collision rates in optical lattices which constrain atomic motion in a periodic potential. Kunugita *et al.* [232] measured a reduced collision rate in a Kr lattice and Lawall *et al.* [237] measured both suppression and enhancement in a Xe lattice. Reference [53] discusses how *very* long-range molecular interactions in a light field might be modified in optical lattices. Finally, we note that Katori *et al.* [211] have measured a quantum statistical effect on ionizing collision rates of metastable Kr. The collision rate for the spin-polarized fermion isotope $^{83}$Kr decreased relative to the collision rate in the unpolarized gas. The decrease was attributed to the different threshold laws for p-waves (which contribute to the spin-polarized case) than for s-waves (which only contribute in the spin-unpolarized case). For a discussion of these threshold laws see Section 2.2.

## 4.3  Ground-state trap-loss collisions

Trap loss processes such as radiative escape and fine-structure changing collisions always start with collision partners approaching on an excited long-range attractive molecular potential. Collisions on the molecular ground state divide into two categories: elastic collisions and hyperfine-changing collisions. Exoergic hyperfine-changing collisions can also lead to loss if the trap is made sufficiently shallow such that the velocity gained by each atom from the HCC kinetic energy release exceeds the maximum capture velocity of the trap.

### Hyperfine-changing collisions

If atoms have nuclear spin $\mathbf{I}$ then the coupling of this nuclear spin to the total electronic angular momentum $\mathbf{J}$ results in hyperfine states $\mathbf{F} = \mathbf{J} + \mathbf{I}$ with projection $M_F$ along some axis of quantization. In binary collisions this axis is always taken to be the line joining the two nuclei with the hyperfine states of the two partners labeled $F_1 M_{F1}$ and $F_2 M_{F2}$. The manifold of molecular hyperfine states $\mathbf{F}_m$ is constructed from vector pairing of the atomic states in the usual way, $\mathbf{F}_m = \mathbf{F}_1 + \mathbf{F}_2$. In the asymptotic limit of internuclear separation the molecular hyperfine states are products of the atomic hyperfine states, $|F_1 M_{F1}\rangle |F_2 M_{F2}\rangle$, and their energy levels are the sum of their constituent atomic hyperfine energies. Figure 2.2. illustrates these molecular potentials near their asymptotic limit for the case of $^{87}$Rb. It is customary to label the asymptotic energy levels by $F_1 F_2$. For example the ground state of Na splits into two hyperfine levels, $F = 1, 2$ by the vector coupling of $J = \frac{1}{2}$ to the nuclear spin $I = \frac{3}{2}$. The energy difference between the two hyperfine levels is 1772 MHz. The molecular hyperfine asymptotes are labeled $F_1 F_2 = 1 + 1$, $1 + 2$, $2 + 2$ with each asymptote separated by 1772 MHz. If during a collisional encounter the exit channel lies on a lower hyperfine level than the entrance channel, the energy difference appears as kinetic energy of the receding partners. In homonuclear collisions each partner receives half the kinetic energy release and an increment in velocity

$$v_{HCC} = \sqrt{\frac{\Delta E_{HCC}}{m}}.$$

If this velocity increment exceeds the capture velocity of the MOT, the hyperfine-changing collision leads to trap loss.

Two physical processes govern the change of hyperfine level: spin exchange and spin dipole–dipole interaction. Spin exchange occurs when the charge clouds of the two colliding partners begin to overlap. The electron spins, $\mathbf{S}_1, \mathbf{S}_2$ decouple from the nuclear spins and recouple to form molecular electron spin states, $\mathbf{S} = \mathbf{S}_1 + \mathbf{S}_2$. The atomic nuclear spins also recouple to form molecular nuclear spin states, $\mathbf{I} = \mathbf{I}_1 + \mathbf{I}_2$ such that $\mathbf{F}_m = \mathbf{S} + \mathbf{I}$ is conserved and $M_F = M_{F1} + M_{F2}$. In the exchange region the asymptotic hyperfine states $|F_1 M_{F1}\rangle |F_2 M_{F2}\rangle$ transform to linear combinations which lead to a finite probability that the postcollisional population of asymptotic states differs from the entrance channel. In contrast to the exchange interaction the long-range spin-dipole interaction is usually weaker by several orders of magnitude, because it derives from relativistic spin-dependent forces proportional to the square of the fine-structure constant. Later we will enlarge the discussion of ground-state collisions and these physical processes in the context of Bose–Einstein condensates (Stoof *et al.* [365]). For the present discussion of trap-loss processes it suffices to note that in a MOT endoergic inelastic channels are generally closed by energy conservation, but the exoergic HCC channels can lead to trap loss.

## Cesium and rubidium

Earlier we noted that Sesko *et al.* [345] observed a rapid increase in $\beta$ as the intensity of their Cs MOT dropped below 4 mW cm$^{-2}$. They interpreted this loss as being due to HCC collisions, but could not infer directly $\beta_{HCC}$, the rate constant for the HCC process itself, because even at the lowest MOT laser intensity the trap loss $\beta$ does not become independent of the trap depth. Sesko *et al.* [345] state a $\beta_{HCC}$ estimation of between $10^{-10}$ and $10^{-11}$ cm$^3$ s$^{-1}$. Wallace *et al.* [422] also observed a rapidly rising branch of $\beta$ as a function of decreasing MOT laser intensity below about 2.5 mW cm$^{-2}$ in a $^{85}$Rb MOT and below about 4 mW cm$^{-2}$ in a $^{87}$Rb MOT. They also reported a strong isotope effect in $\beta$ and showed that the ratio of $\beta_{85}/\beta_{87}$ at constant trap intensity supported the view that MOT velocity-dependent forces rather than position-dependent forces dominate atom capture and trapping in the MOT environment. Furthermore, at the lowest MOT intensities $\beta$ shows some sign of becoming independent of MOT capture velocity, which would imply a direct measure of $\beta_{HCC}$. Wallace *et al.* [422] estimated this $\beta_{HCC}$ to be about the same for both isotopes, $\sim 2 \times 10^{-11}$ cm$^3$ s$^{-1}$. If MOTs functioned as spatially isotropic, conservative potentials, then one would expect a very sharp threshold for loss as the trapping laser intensity decreases below the minimum required to contain the HCC kinetic energy release. In fact, to first approximation, the MOT restoring forces are the sum of a velocity-dependent dissipative term and a spatially dependent term, so it is more accurate to characterize a MOT by its capture velocity rather than the trap "depth". This capture velocity is a complicated function of trapping laser detuning, intensity, spot size, magnetic field gradient; and, as Ritchie *et al.* [328] have shown in detail for a Li MOT, exhibits spatial anisotropy. Therefore, as the MOT laser intensity decreases, the trap will start to leak in the shallowest directions (along the MOT symmetry axis for a six-beam MOT) first. As the MOT laser intensity continues to decrease, the leak rate will increase until $\beta$ reaches a constant value where atoms are no longer contained in any direction. Ritchie *et al.* [328, 329] have shown how a detailed trap simulation can be used to extract $\beta_{FCC}$ from the intensity dependence of MOT performance for fine-structure-changing collisions in Li, but a similar analysis to extract $\beta_{HCC}$ has not yet been attempted.

## Sodium

Marcassa *et al.* [257] could not maintain their Na MOT at intensities sufficiently low to measure $\beta$ on the rapidly rising branch where HCC is thought to become the dominant trap-loss process. Although similar HCC losses have been observed in Cs (Sesko *et al.* [345]), Rb (Wallace *et al.* [422], and K (Santos *et al.* [340]), the low-intensity branch is more difficult to observe in Na because the threshold for its occurrence starts at MOT intensities already weak compared to the saturation intensity of the atomic cooling transition. At such low intensities the vapor-loaded MOT barely functions. However, starting from a slow atomic beam Shang *et al.* [349] succeeded in efficiently loading an Na MOT to investigate trap losses over a wide range of MOT intensity, including the low-intensity $\beta$ branch. Another novel aspect of their work is the investigation of trap loss with all ground-state trapped atoms prepared in

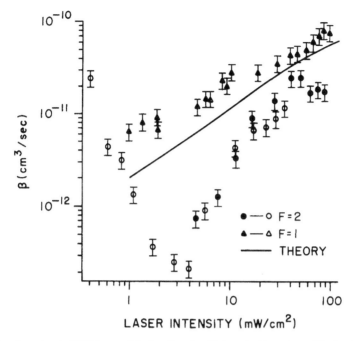

**Fig. 4.36.** Trap loss due to HCC is possible for $F = 2$ collisions but is not possible for $F = 1$ collisions. The absence of a rapidly rising branch at low intensity for $F = 1$ collisions confirms the trap loss process at low trap depth, from Shang *et al.* [349].

*either* the $3s\,^2S_{1/2}(F = 1)$ hyperfine level, *or* the $3s\,^2S_{1/2}(F = 2)$ level. The motivation was to show directly that, since only collisions between atoms prepared in the $F = 2$ level can lead to HCC, only they exhibit the low-intensity $\beta$ branch. Their trap-loss measurements presented in Fig. 4.36 confirm that $F = 1$ collisions do not exhibit the rapidly rising $\beta$ signature of HCC. In contrast, for atoms trapped in the $F = 2$ ground state, Shang *et al.* [349] did observe the conventional decrease of $\beta$ as the MOT intensity decreases, until a minimum is reached at about $4\,\mathrm{mW\,cm}^{-2}$. Below $4\,\mathrm{mW\,cm}^{-2}$ HCC dominates, and the loss rate increases rapidly. For atoms in the $F = 1$ hyperfine level the HCC exoergic channel is closed, and $\beta$ exhibits no rapidly rising branch below the $4\,\mathrm{mW\,cm}^{-2}$ minimum. To obtain a trap with atoms predominantly in the $F = 1$ ground hyperfine level, Shang *et al.* [349] used the type II trap described by Prentiss *et al.* [315] and later used by Marcassa *et al.* [257]. The values measured by Shang *et al.* [349] are in reasonable agreement with theory presented in Marcassa *et al.* [257] in which $\beta$ from FCC and RE were calculated using the OBE approach of [32] and accurate $Na_2$ potential curves from Magnier *et al.* [254]. We have previously discussed the limitations of the OBE approach in Section 4.1.5. The solid line in Fig. 4.36 shows the theory calculation. Comparison between $\beta_{F1}$ and $\beta_{F2}$ on the high-intensity branch shows that trap loss rate from $F = 1$ collisions is greater than $F = 2$ collisions by about a factor of 3 over the intensity range from 20 to 90 $\mathrm{mW\,cm}^{-2}$. From about 4 to 20 $\mathrm{mW\,cm}^{-2}$ $\beta_{F2}$ drops by about two orders of magnitude from $\sim 2 \times 10^{-10}$ to

$\sim 2 \times 10^{-12}\,\mathrm{cm}^3\,\mathrm{s}^{-1}$. Shang *et al.* [349] speculate that sub-Doppler cooling in this intensity range may slow the collision rate. Differences in molecular hyperfine structure between $F = 2$ collisions and $F = 1$ collisions may also play a role. At present the dramatic drop in $\beta_{F2}$ is not clearly explained.

The most recent contribution to ground-state trap loss in sodium corresponds to a study of the collisional loss rate of Na in a MOT operating on the D1-line (Marcassa *et al.* [258]). Although the idea of trapping alkalis using the D1 resonance is straightforward, only recently has it been demonstrated by Flemming *et al.* [136]. The performance of a D1 trap alone in a vapor cell is very poor and requires loading from an external source of cold atoms. Therefore, Marcassa *et al.* [258] used a conventional MOT operating on the D2-line for loading the D1 MOT. Atoms are loaded by capturing and cooling them in a conventional D2 MOT with the D1 optical beams superposed in the MOT configuration. After loading, the D2-line laser is turned off, but the cold atoms remain trapped in the D1 MOT. The number of trapped atoms is determined during the D2 MOT. As previously, $\beta_{D1}$ is extracted from the fluorescence decay during the D1-line cycle. The measured $\beta_{D1}$ from several intensities of laser at a detuning of about $-10\,\mathrm{MHz}$ from $3S_{1/2}(F = 2) \rightarrow 3P_{1/2}(F = 2)$ (and with the repumper close to $3S_{1/2}(F = 1) \rightarrow 3P_{1/2}(F = 2)$) is presented in Fig. 4.37. In the range from 200 to about $65\,\mathrm{mW\,cm}^{-2}$ no significant variation in $\beta$ is observed. At about $65\,\mathrm{mW\,cm}^{-2}$ the collisional loss rate increases considerably. At this point of rapid increase, the trap has become shallow enough to make HCC an important loss mechanism. For a trap operating at the D2-line the intensity at which this rapid increase occurs is about $5\,\mathrm{mW\,cm}^{-2}$ (Shang *et al.* [349]), about 10 times smaller than for the D1-line trap. As a consequence the trap depth of the D1-line trap is about $1/10$ of the D2-line trap at comparable intensities. This considerable decrease in trap depth is attributed to the several possible "dark states" that can occur in the D1-line trap.

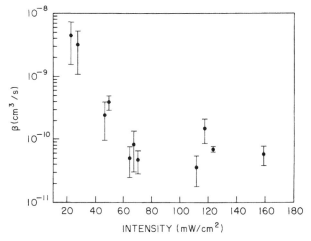

**Fig. 4.37.** Trap loss from an Na MOT formed on the D1 resonance line as a function of MOT laser intensity (trap depth), from Marcassa *et al.* [258].

### Summary comparison of trap loss rate constants

Many research groups have carried out HCC trap loss experiments in MOTs with varied instrumental design parameters, different MOT laser intensities, spot sizes, and detunings, and a host of atoms all with their own resonance lines and lifetimes. It would be desirable to organize these measurements on a common basis so as to identify the critical factors resulting in a particular determination of $\beta$. A simple model of MOT atom confinement from which a parameter $\Theta$ is derived that correlates well with the measurements discussed in this chapter is presented in Weiner *et al.* [441].

It is assumed that an atom pair subject to an HCC collision receives an extra quantity of energy $\Delta E_{HCC}$ corresponding to the exoergicity of the transition and that each atom of mass $m$ carries away half the velocity increase,

$$v = \left( \frac{\Delta E_{HCC}}{m} \right)^{1/2}.$$  (4.36)

It is assumed further that the atom travels outward from the center of the trap to the radius of the MOT $w$ subject only to the dissipative velocity-dependent decelerating force. Wallace *et al.* [422] cite evidence from their isotopic studies in an Rb MOT that the velocity-dependent force dominates over the position-dependent restoring force. Thus

$$F = -\alpha v,$$  (4.37)

and

$$v(z) = \left( \frac{\Delta E_{HCC}}{m} \right)^{1/2} - \frac{\alpha}{m} z.$$  (4.38)

The maximum capture velocity is defined by

$$v(w) = 0,$$  (4.39)

so

$$\frac{\alpha}{m} w = \left( \frac{\Delta E_{HCC}}{m} \right)^{1/2} = v_{max}.$$  (4.40)

Now from Eq. 3.22,

$$\alpha = \frac{16 \hbar k^2 \Omega^2 \Delta \Gamma}{\left( 4\Delta^2 + \Gamma^2 + 2\Omega^2 \right)^2},$$  (4.41)

so at relatively large detunings and low intensities the second two terms in the denominator of Eq. 4.41 can be ignored and writing the MOT intensity $I$ in terms of the Rabi frequency $\Omega$,

$$\Omega^2 = \frac{I}{I_{sat}} \Gamma^2,$$  (4.42)

Weiner *et al.* [441] find

$$v_{max} = \frac{\hbar k^2 I w \Gamma^3}{\Delta^3 I_{sat} m}.$$  (4.43)

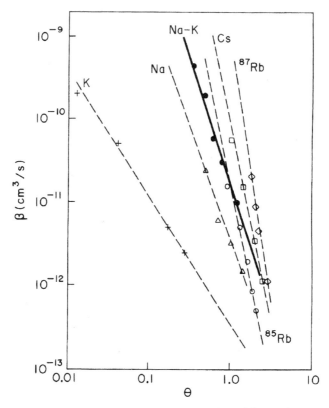

**Fig. 4.38.** Summary plot of the trap loss constant $\beta$ as a function of the trap loss parameter $\Theta$ for all the alkali systems studied to date. From [441].

Removing all the constant terms a trap capture parameter $\Theta$ is defined as

$$\Theta = \frac{1}{\lambda^2} \frac{\Gamma^3}{\Delta^3} \frac{I}{I_{sat}} \frac{w}{m}. \tag{4.44}$$

Figure 4.38 shows $\beta$ plotted against $\Theta$ for all the alkalis discussed above. Note that for heteronuclear collisions Eq. 4.44 must be modified by substituting $\frac{m_1}{2}(1 + \frac{m_1}{m_2})$ for $m$ where $m_1$ is the mass of the species of the observed trap loss.

### 4.3.1   Low-intensity trap loss revisited

Since the early measurements of Sesko *et al.* on trap loss in cesium [345] or Wallace *et al.* in rubidium [422], it has become conventional wisdom to interpret the rising branch of the trap-loss rate constant $\beta$ as being due to energy release from hyperfine changing collisions (see for example Fig. 4.17 or Fig. 4.19). According to this interpretation the low-intensity rising branch should level off as the trap intensity continues to decrease. The reason for this is that the trap depth is proportional to the intensity. At some low-intensity threshold the kinetic energy released by the HCC process should be sufficient to overcome

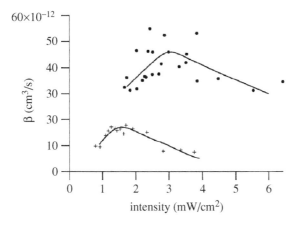

**Fig. 4.39.** Loss rates from the measurements reported in [304] showing the unexpected decrease in $\beta$ at very low intensity. The crosses and filled circles show data taken at trap detunings of $-1.5\Gamma$ and $-2\Gamma$, repectively.

the shallow trap depth, and the loss rate constant would become independent of further trap intensity decrease. This interpretation was called into question when Nesnidal and Walker [304] reported measurements of the trap loss rate constant in a $^{87}$Rb MOT that clearly *decreased* below a trap intensity of a few mW cm$^{-2}$. Figure 4.39 shows an example of their data. Nesnidal and Walker concluded that the loss rate must indeed be coming from ground-state collisions, but that the HCC process must somehow be light-intensity dependent. They pointed out that in the special case of $^{87}$Rb collisions, the conventional spin-exchange probability (i.e. the process taking place in the absence of an external light field) is orders of magnitude smaller than expected. The reason is that this rate constant must be proportional to $\sin^2(\eta_s - \eta_t)$, where $\eta_s$ and $\eta_t$ are scattering phase shifts for the singlet and triplet potentials, and quite coincidentally these two phase shifts are nearly numerically identical. Therefore in the special case of $^{87}$Rb collisions, in the absence of light, HCC is suppressed by a destructive interference of the scattering from the singlet and triplet ground-state potentials. Any perturbation inducing a greater difference in the phase shifts would tend to enhance the HCC rate constant. Nesnidal and Walker therefore suggest that the light field of the MOT may be increasing the phase difference, but they offer no detailed model that would explain the observed sensitivity to MOT detuning or the magnetic field gradient.

Telles *et al.* [380] have suggested an alternative explanation that would apply to all low-intensity trap loss collisions, not just the special case of $^{87}$Rb. Telles *et al.* suggest that the low-intensity peak in $\beta$ arises from the light-intensity dependence of the radiative escape process and not from any spin exchange interaction. Figure 4.40 shows the measured trap loss rate constant from the data of [304] and the results of the Telles *et al.* model. Despite the fact that the model does not agree quantitatively with the experimental measurements, the form and the position of the minimum in $\beta$ are well reproduced for most of the alkali collisions. However this modified radiative escape model cannot be the whole story. If HCC

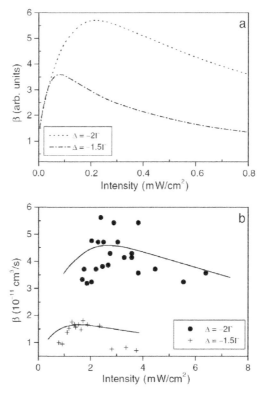

**Fig. 4.40.** (a) Calculated trap loss rates for two different MOT laser detunings: $\Delta = -2\Gamma$ (dotted line) and $\Delta = -1.5\Gamma$ (dash-dotted line). (b) Comparison of experimental results from [304] and the calculation of [380]. The quantitative results of the model have been rescaled to best fit the experimental points.

is not important, it is difficult to explain the results of [349] (see Fig. 4.36), which show that in the case of sodium collisions, the low-intensity branch was observed for $Na(F = 2)$ collisions but not for $Na(F = 1)$ collisions. Clearly in this case at least HCC must play an important role. The open question is therefore: does $^{87}Rb$ present a special case of peculiar quantum destructive interference or is the low-intensity trap loss peak a general property of the RE mechanism at low MOT intensity? This question once again illustrates the unsatisfactory state of understanding for trap loss collisions in the presence of optical fields within a few line widths of the resonance transition.

# 5

# Photoassociation spectroscopy

## 5.1  Introduction

If, while approaching on an unbound ground-state potential, two atoms absorb a photon and couple to an excited bound molecular state, they are said to undergo photoassociation. Figure 5.1 illustrates the process. At long range electrostatic dispersion forces give rise to the ground-state molecular potential varying as $C_6/R^6$. If the two atoms are homonuclear, then a resonant dipole–dipole interaction sets up $\pm C_3/R^3$ excited-state repulsive and attractive potentials. Figure 5.2 shows the actual long-range excited potential curves for the sodium dimer, originating from the $^2S_{1/2} + {}^2P_{3/2}$ and $^2S_{1/2} + {}^2P_{1/2}$ separated atom states. For cold and ultracold photoassociation processes the long-range attractive potentials play the key role; the repulsive potentials figure importantly in optical shielding and suppression, the subject of Chapter 6. In the presence of a photon with frequency $\omega_p$ the colliding pair with kinetic energy $k_B T$ couples from the ground-state to the attractive molecular state in a free–bound transition near the Condon point $R_C$, the point at which the difference potential just matches $\hbar\omega_p$.

Scanning the probe laser $\omega_p$ excites population of vibration–rotation states in the excited bound potential and generates a free–bound spectrum. This general class of measurements is called photoassociative spectroscopy (PAS) and can be observed in several different ways. The observation may consist of bound–state decay by spontaneous emission, most probably as the nuclei move slowly around the outer turning point, to some distribution of continuum states on the ground potential controlled by bound–free nuclear Franck–Condon overlap factors. The free atoms then recede on the ground-state with some distribution of kinetic energy. In a MOT or a FORT, this radiative escape (RE) process can be an important source of loss if each atom receives enough kinetic energy to overcome trap restoring forces. As $\omega_p$ passes through the free–bound resonances, the loss channel opens; and the number of atoms in the trap sharply diminishes. Thus scanning $\omega_p$ generates a high-resolution fluorescence-loss spectrum. The excited bound quasimolecule may also decay around the inner turning point to bound levels of the ground molecular state. Since the stable dimers are usually not confined by MOT fields, this cold-molecule production is another trap-loss channel, although the branching ratio to ground-state bound molecules is usually small, only a few per cent of the total loss rate.

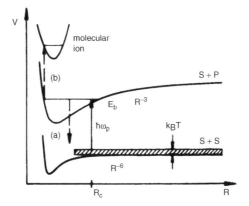

**Fig. 5.1.** Diagram of the photoassociation process. First excitation ($\hbar\omega_p$) is followed by (a) spontaneous or stimulated emission back to bound or continuum levels of the ground-state or (b) second excitation to molecular ion (photoassociative ionization).

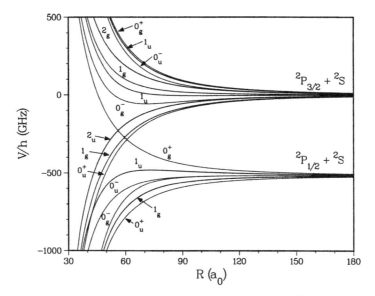

**Fig. 5.2.** Long-range potential curves of the first excited states of the Na dimer labeled by Hund's case (c) notation. The number gives the projection of electron spin plus orbital angular momentum on the molecular axis. The g/u and $+/-$ labels correspond to conventional point-group symmetry notation.

Instead of observing the spectrum by spontaneous decay and trap loss, a second probe photon may excite the long-range bound quasimolecule to a higher excited state (or manifold of states) from which the molecule may undergo direct photoionization or autoionization within some range of internuclear separation. If this second photon comes from the same light source as the first, the process is called "one-color" PAS, and if from an independently tunable source then "two-color" PAS. When the ultimate product of the photoassociation is a molecular ion, the process is called photoassociative ionization (PAI).

## 5.2 Photoassociation at ambient and cold temperatures

The first measurement of a free–bound photoassociative absorption appeared long before the development of optical cooling and trapping, about two decades ago, when Scheingraber and Vidal [341] reported the observation of photoassociation in collisions between magnesium atoms. In this experiment fixed ultraviolet emission lines from an argon ion laser excited free–bound transitions from the thermal continuum population of the ground $X\,^1\Sigma_g^+$ state to bound levels of the $A\,^1\Sigma_u^+$ state of $Mg_2$. Scheingraber and Vidal analyzed the subsequent fluorescence to bound and continuum states from which they inferred the photoassociative process. The first unambiguous photoassociation *excitation spectrum*, however, was measured by Inoue *et al.* [186] in collisions between Xe and Cl at 300 K. In both of these early experiments the excitation was not very selective due to the broad thermal distribution of populated continuum ground-states. Jones *et al.* [197] with a technically much improved experiment reported beautiful free–bound vibration progressions in KrF and XeI X → B transitions; and, from the intensity envelope modulation were able to extract the functional dependence of the transition moment on the internuclear separation. Although individual vibrational levels of the *B* state were clearly resolved, the underlying rotational manifolds were not. Jones *et al.* [197] simulated the photoassociation structure and line shapes by assuming a thermal distribution of rotational levels at 300 K. Photoassociation and dissociation processes prior to the cold and ultracold epoch have been reviewed by Tellinghuisen [383].

A decade after Schenigraber and Vidal reported the first observation of photoassociation Thorsheim *et al.* [385] proposed that high-resolution free–bound molecular spectroscopy should be possible using optically cooled and confined atoms. Figure 5.3 shows a portion of their calculated X → A absorption spectrum at 10 mK for sodium atoms. This figure illustrates how cold temperatures compress the Maxwell–Boltzmann distribution to the point where individual rotational transitions in the free–bound absorption are clearly resolvable. The marked differences in peak intensities indicate scattering resonances, and the asymmetry in the line shapes, tailing off to the red, reflect the thermal distribution of ground-state collision energies at 10 mK. Figure 5.4 plots the photon-flux-normalized absorption rate coefficient for singlet $X\,^1\Sigma_g^+ \rightarrow A\,^1\Sigma_u^+$ and triplet $a\,^3\Sigma_u^+ \rightarrow 1\,^3\Sigma_g^+$ molecular transitions over a broad range of photon excitation, red-detuned from the Na($^2$S → $^2$P) atomic resonance line. The strongly modulated intensity envelopes are called Condon modulations, and they reflect the overlap between the ground-state continuum wavefunctions and the bound excited vibrational wavefunctions. These can be understood from the reflection approximation described in Eq. 2.57. We shall see later that these Condon modulations reveal detailed information about the ground-state scattering wavefunction and potential from which accurate s-wave scattering lengths can be determined. Thorsheim *et al.* [385], therefore, predicted three of the principal features of ultracold photoassociation spectroscopy later to be exploited in many follow-up experiments: (1) precision measurement of vibration–rotation

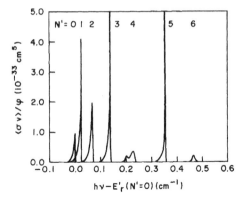

**Fig. 5.3.** Calculated free–bound photoassociation spectrum at 10 mK from Thorsheim *et al.* [385].

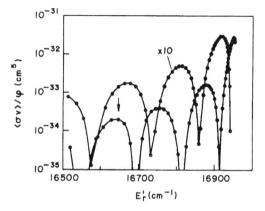

**Fig. 5.4.** Calculated absorption spectrum of photoassociation in Na at 10 mK, showing Condon fluctuations. The curve labeled "×10" refers to $^3\Sigma_g \rightarrow {}^3\Sigma_u$, while the other curve refers to $^1\Sigma_g \rightarrow {}^1\Sigma_u$ transitions, from Thorsheim *et al.* [385].

progressions from which accurate excited-state potential parameters can be determined; (2) line profile measurements and analysis to determine collision temperature and threshold behavior; and (3) spectral intensity modulation from which the ground-state potential, the scattering wavefunction, and the s-wave scattering length can be characterized with great accuracy.

An important difference distinguishes ambient temperature photoassociation in rare-gas halide systems and sub-millikelvin temperature photoassociation in cooled and confined alkali systems. At temperatures found in MOTs and FORTs (and within selected velocity groups or intra-beam collisions in atomic beams) the collision dynamics are controlled by long-range electrostatic interactions, and Condon points $R_C$ are typically at tens to hundreds of $a_0$. In the case of the rare-gas halides the Condon points are in the short-range region of chemical binding, and therefore free–bound transitions take place at much smaller internuclear distances, typically less than ten $a_0$. For the colliding A, B quasimolecule the pair density $n$ as a function of $R$ is given by

$$n = n_A n_B 4\pi R^2 e^{-V(R)/kT} \tag{5.1}$$

so the density of pairs varies as the square of the internuclear separation. Although the pair-density $R$ dependence favors long-range photoassociation, the atomic reactant densities are quite different with the $n_A n_B$ product on the order of $10^{35}$ cm$^{-6}$ for rare-gas halide photoassociation and only about $10^{22}$ cm$^{-6}$ for optically trapped atoms. Therefore the effective pair density available for rare-gas halide photoassociation greatly exceeds that for cold alkali photoassociation, permitting fluorescence detection and dispersion by high resolution (but inefficient) monochromators.

## 5.3 Associative and photoassociative ionization

Conventional associative ionization (AI) occurring at ambient temperature proceeds in two steps: excitation of isolated atoms followed by molecular autoionization as the two atoms approach on excited molecular potentials. In sodium for example (Weiner [439]),

$$Na + \hbar\omega \rightarrow Na^* \tag{5.2}$$
$$Na^* + Na^* \rightarrow Na_2^+ + e. \tag{5.3}$$

The collision event lasts a few picoseconds, which is fast compared to radiative relaxation of the excited atomic states ($\sim$ tens of nanoseconds). Therefore the incoming atomic excited states can be treated as stationary states of the system Hamiltonian, and spontaneous radiative loss does not play a significant role. In contrast, cold and ultracold photoassociative ionization must always start on ground-states because the atoms move so slowly that radiative lifetimes become short compared to the collision duration. The partners must be close enough at the Condon point, where the initial photon absorption takes place, so that a significant fraction of the excited scattering flux survives radiative relaxation and goes on to populate the final inelastic channel. Thus PAI is also a two-step process: (1) photoexcitation of the incoming scattering flux from the molecular ground-state continuum to specific vibration–rotation levels of a bound molecular state; and (2) subsequent photon excitation either to a doubly-excited molecular autoionizing state or directly to the molecular photoionization continuum. For example in the case of sodium collisions the principal route is through doubly excited autoionization (Julienne and Heather, [201]),

$$Na + Na + \hbar\omega_1 \rightarrow Na_2^* + \hbar\omega_2 \rightarrow Na_2^{**} \rightarrow Na_2^+ + e. \tag{5.4}$$

whereas for rubidium atoms the only available route is direct photoionization in the second step (Leonhardt and Weiner [241]),

$$Rb + Rb + \hbar\omega_1 \rightarrow Rb_2^* + \hbar\omega_2 \rightarrow Rb_2^+ + e. \tag{5.5}$$

Collisional ionization can play an important role in plasmas, flames, atmospheric and interstellar physics and chemistry. Models of these phenomena depend critically on the accurate determination of absolute cross sections and rate coefficients. The rate coefficient is the quantity closest to what an experiment actually measures and can be regarded as the cross section averaged over the collision velocity distribution,

$$K = \int_0^\infty v\sigma(v)f(v)\,\mathrm{d}v. \tag{5.6}$$

The velocity distribution $f(v)$ depends on the conditions of the experiment. In cell and trap experiments it is usually a Maxwell–Boltzmann distribution at some well-defined temperature, but $f(v)$ in atomic beam experiments, arising from optical excitation velocity selection, deviates radically from the normal thermal distribution [324, 402, 403]. The actual signal count rate, $\mathrm{d}(X_2^+)/\mathrm{d}t$, relates to the rate coefficient through

$$\frac{1}{V\alpha}\frac{\mathrm{d}(X_2^+)}{\mathrm{d}t} = K[X]^2, \tag{5.7}$$

where $V$ is the interaction volume, $\alpha$ is the ion detection efficiency, and $[X]$ is the atom density. If rate constant or cross section measurements are carried out in crossed or single atomic beams (Weiner et al. [442], Thorsheim et al. [384], Tsao et al. [403]) special care is necessary to determine the interaction volume and atomic density. PAI was the first measured collisional process observed between cooled and trapped atoms (Gould et al. [158]). The experiment was performed with atomic sodium confined in a hybrid laser trap, utilizing both the spontaneous radiation pressure and the dipole force. The trap had two counter-propagating, circularly polarized Gaussian laser beams brought to separate foci such that longitudinal confinement along the beam axis was achieved by the spontaneous force and transversal confinement by the dipole force. The trap was embedded in a large ($\sim 1$ cm diameter) conventional optical molasses loaded from a slowed atomic beam. The two focused laser beams comprising the dipole trap were alternately chopped with a 3 µs "trap cycle", to avoid standing-wave heating. This trap cycle for each beam was interspersed with a 3 µs "molasses cycle" to keep the atoms cold. The trap beams were detuned about 700 MHz to the red of the $3s\,^2S_{1/2}(F=2) \to 3p\,^2P_{3/2}(F=3)$ transition while the molasses was detuned only about one natural line width ($\sim 10$ MHz). The atoms captured from the molasses ($\sim 10^7$ cm$^{-3}$) were compressed to a much higher excited atom density ($\sim 5 \times 10^9$ cm$^{-3}$) in the trap. The temperature was measured to be about 750 µK. Ions formed in the trap were accelerated and focused toward a charged-particle detector. To ensure the identity of the counted ions, Gould et al. [158] carried out a time-of-flight measurement; the results of which, shown in Fig. 5.5, clearly establish the Na$_2^+$ ion product. The linearity of the ion rate with the square of the atomic density in the trap, supported the view that the detected Na$_2^+$ ions were produced in a binary collision. After careful measurement of ion rate, trap volume and excited atom density, the value for the rate coefficient was determined to be $K = (1.1^{+1.3}_{-0.5}) \times 10^{-11}$ cm$^3$ s$^{-1}$. Gould et al. [158], following conventional wisdom, interpreted the ion production as coming from collisions between two *excited* atoms,

$$\frac{\mathrm{d}N_I}{\mathrm{d}t} = K\int n_e^2(\mathbf{r})\,\mathrm{d}^3\mathbf{r} = K\,\hat{n}_e\,N_e, \tag{5.8}$$

where $\mathrm{d}N_I/\mathrm{d}t$ is the ion production rate, $n_e(\mathbf{r})$ is the excited-state density, $N_e$ is the number of excited atoms in the trap ($= \int n_e(\mathbf{r})\,\mathrm{d}^3\mathbf{r}$), and $\hat{n}_e$ is the "effective" excited-state trap

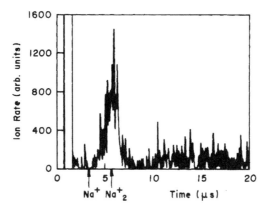

**Fig. 5.5.** Time-of-flight spectrum clearly showing that the ions detected are $Na_2^+$ and not the atomic ion, from Gould *et al.* [158].

density. The value for $K$ was then determined from these measured parameters. Assuming an average collision velocity of 130 cm s$^{-1}$, equivalent to a trap temperature of 750 μK, the corresponding cross section was determined to be $\sigma = (8.6^{+10.0}_{-3.8}) \times 10^{-14}$ cm$^2$. In contrast, the cross section at $\sim$575 K had been previously determined to be $\sim 1.5 \times 10^{-16}$ cm$^2$ (Bonnano *et al.* [55]; Wang *et al.* [434]; Wang *et al.* [435]). Gould *et al.* [158] rationalized the difference in cross section size by invoking the difference in de Broglie wavelengths, the number of participating partial waves, and the temperature dependence of the ionization channel probability. The quantal expression for the cross section in terms of partial wave contributions $l$ and inelastic scattering probability $S_{12}$ is

$$\sigma_{12}(\epsilon) = \left(\frac{\pi}{k^2}\right) \sum_{l=0}^{\infty} (2l+1)|S_{12}(\epsilon, l)|^2 \tag{5.9}$$

$$\cong \pi \left(\frac{\lambda_{dB}}{2\pi}\right)^2 (l_{max}+1)^2 P_{12},$$

where $\lambda_{dB}$ is the entrance channel de Broglie wavelength and $P_{12}$ is the probability of the ionizing collision channel averaged over all contributing partial waves of which $l_{max}$ is the greatest. The ratio of the $(l_{max}+1)^2$ between 575 K and 750 μK is about 400 and the de Broglie wavelength ratio factor varies inversely with temperature. Therefore in order that the cross section ratio to be consistent with low- and high-temperature experiments, $\frac{\sigma_{12}(575K)}{\sigma_{12}(750\mu K)} \sim 1.7 \times 10^{-3}$, Gould *et al.* [158] concluded that $P_{12}$ must be about three times greater at 575 K than at 750 μK. However, it soon became clear that the conventional picture of associative ionization, starting from the excited atomic states, could not be appropriate in the cold regime. Julienne [198] pointed out the essential problem with this picture. In the molasses cycle the optical field is only red-detuned by one line width, and the atoms must therefore be excited at very long range, near 1800 a$_0$. The collision travel time to the close internuclear separation where associative ionization takes place is long compared to the radiative lifetime, and most of the population decays to the ground-state before reaching the autoionization zone. During the trap cycle, however, the excitation takes place at much

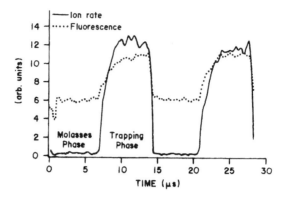

**Fig. 5.6.** Trap modulation experiment showing much greater depth of ion intensity modulation (by more than one order of magnitude) than fluorescence or atom number modulation, demonstrating that excited atoms are not the origin of the associative ionizing collisions, Lett *et al.* [244].

closer internuclear distances due to a 70 line width red detuning and high-intensity field dressing. Therefore one might expect excitation survival to be better during the trap cycle than during the molasses cycle, and the NIST group, led by W. D. Phillips, set up an experiment to test the predicted cycle dependence of the ion rate.

Lett *et al.* [244] performed a new experiment using the same hybrid trap. This time, however, the experiment measured ion rates and fluorescence separately as the hybrid trap oscillated between "trap" and "molasses" cycles. The results from this experiment are shown in Fig. 5.6. While keeping the total number and density of atoms (excited atoms plus ground-state atoms) essentially the same over the two cycles and while the excited-state *fraction* changed only by about a factor of 2, the ion rate increased in the trapping cycle by factors ranging from 20 to 200 with most observations falling between 40 and 100. This verified the predicted effect qualitatively even if the magnitude was smaller than the estimated $10^4$ factor of Julienne [198]. This modulation ratio is orders of magnitude more than would be expected if excited atoms were the origin of the associative ionization signal. Furthermore, by detuning the trapping lasers over 4 GHz to the red, Lett *et al.* [244] continued to measure ion production at rates comparable to those measured near the atomic resonance. At such large detunings reduction in atomic excited-state population would have led to reductions in ion rate by over four orders of magnitude had the excited atoms been the origin of the collisional ionization. Not only did far off-resonance trap cycle detuning maintain the ion production rate, Lett *et al.* [244] observed evidence of peak structure in the ion signal as the dipole trap cycle detuned to the red.

To interpret the experiment of Lett *et al.* [244], Julienne and Heather [201] proposed a mechanism which has become the standard picture for cold and ultracold photoassociative ionization. Figure 5.7 details the model. Two colliding atoms approach on the molecular ground-state potential. During the molasses cycle with the optical fields detuned only about one line width to the red of atomic resonance, the initial excitation occurs at very long range, around a Condon point at 1800 $a_0$. A second Condon point at 1000 $a_0$ takes the population to a $1_u$ doubly excited potential that, at shorter internuclear distance, joins

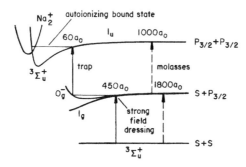

**Fig. 5.7.** Photoassociative ionization in Na collisions, from Heather and Julienne [166].

adiabatically to a $^3\Sigma_u^+$ potential, thought to be the principle short-range entrance channel to associative ionization (Dulieu et al. [121], Henriet *et al.* [169]). More recent calculations suggest other entrance channels are important as well [119]. The long-range optical coupling to excited potentials in regions with little curvature implies that spontaneous radiative relaxation will depopulate these channels before the approaching partners reach the region of small internuclear separation where associative ionization takes place. The overall probability for collisional ionization during the molasses cycle remains therefore quite low. In contrast, during the trap cycle the optical fields are detuned 60 line widths to the red of resonance, the first Condon point occurs at 450 $a_0$; and, if the trap cycle field couples to the $0_g^-$ long-range molecular state (Stwalley *et al.* [368]), the second Condon point occurs at 60 $a_0$. Survival against radiative relaxation improves greatly because the optical coupling occurs at a much shorter range where excited-state potential curvature accelerates the two atoms together. Julienne and Heather [201] calculated about a three-order-of-magnitude enhancement in the rate constant for collisional ionization during the trap cycle. The dashed and sold arrows in Fig. 5.7 indicate the molasses cycle and trap cycle pathways, respectively. The strong collisional ionization rate constant enhancement in the trap cycle calculated by Julienne and Heather [201] is roughly consistent with the measurements of Lett *et al.* [244], although the calculated modulation ratio is somewhat greater than what was actually observed. Furthermore, Julienne and Heather calculate structure in the trap detuning spectrum. As the optical fields in the dipole trap tune to the red, a rather congested series of ion peaks appear which Julienne and Heather ascribed to free–bound association resonances corresponding to vibration–rotation bound levels in the $1_g$ or the $0_g^-$ molecular excited states. The density of peaks corresponded roughly to what Lett *et al.* [244] had observed; these two tentative findings together were the first evidence of a new photoassociation spectroscopy. In a subsequent full paper expanding on their earlier report, Heather and Julienne [166] introduced the term, "photoassociative ionization" to distinguish the two-step optical excitation of the quasimolecule from the conventional associative ionization collision between excited atomic states. In a very recent paper, P. Pillet *et al.* [314] have developed a perturbative quantum approach to the theory of photoassociation which can be applied to the whole family of alkali homonuclear molecules. This study presents a useful table of photoassociation rates which reveals an important trend toward lower rates of molecule formation as

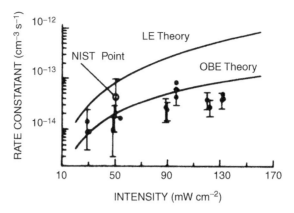

**Fig. 5.8.** Photoassociative ionization rate constant as a function of MOT light intensity. MOT laser detuned to the red by approximately $\Gamma$, from [25].

the alkali mass increases and provides a helpful guide to experiments designed to detect ultracold molecule production.

### 5.3.1 PAI at small detuning

When near-resonant light couples ground and excited molecular states of colliding atoms, spontaneous emission occurs during the collision with high probability. Collision problems involving strong coupling to one or more optical field modes and with radiative dissipation can only be properly treated by solving the Liouville equation of motion for the quantum density matrix of particles plus field,

$$\frac{d}{dt}\rho(t) = \frac{1}{i\hbar}\,[H, \rho(t)]\,, \tag{5.10}$$

where $\rho(t)$ is the density matrix describing the states of the quasimolecule and the radiation field and $H = H_{molec} + H_{field} + H_{int}$ is the Hamiltonian for the system of quasimolecule, radiation field, and the coupling between the two. Photoassociative ionization during the molasses cycle of the NIST hybrid trap or within the confines of a magneto-optical trap correspond to this situation. Reference [25], therefore, undertook to investigate the PAI rate constant in a sodium MOT as a function of MOT optical field intensity. The essential idea was to measure the dependence of the PAI rate constant on MOT intensity and compare it to rate constants calculated from an optical-Bloch-equation (OBE) theory, based on Eq. 5.10, proposed earlier by [32]. Reference [25] also compared the experimental results to the Gallagher–Pritchard model or "local equilibrium," (LE) model discussed at length in Chapter 4 on exoergic trap loss. The results are shown in Fig. 5.8. The semiclassical OBE theory agrees well with the experiment at the low end of the intensity scale, but the measured rate constants appear to increase less rapidly with increasing intensity than the calculations and to flatten around $100\ \mathrm{mW\ cm^{-2}}$. The LE model builds on the quasistatic theory of collision broadening, treats the molecule–field interaction perturbatively, and does not attempt to solve a density matrix equation of motion. The term "local equilibrium" comes from the characteristic of this theory that at any Condon point, the detuning dependence of the optical excitation takes the form of a Lorentzian line shape. The key assumption, therefore, of the LE model is that at any internuclear separation the quasimolecule has time to reach "equilibrium" with the modes of the radiation field. In both the OBE and LE models, calculation of the rate constant requires determining two probabilities: first the probability that a two-step optical excitation-survival process (first to the $1_g$ or $0_g^-$, and then to the $1_u$) brings the two atoms together near $R \sim 6\,a_0$ on the doubly excited autoionizing $^3\Sigma_u^+$ potential; secondly, the probability that the autoionization takes place. Both models require the molecular autoionization probability and rely on measurements reported in the literature (Huennekens and Gallagher [183]; [55]). The two models differ significantly in the calculation of the first step, the optical excitation probability. Figure 5.8 shows that the LE theory overestimates the PAI rates by about one order of magnitude. Neither theory takes into account the extensive molecular hyperfine structure of the excited states, and the evidence from trap loss experiments (see Chapter 4) indicates that models of collision dynamics at small detuning which ignore molecular excited-state hyperfine structure usually fail.

Recently, Blangé *et al.* [45] have performed photodissociation fragmentation spectroscopy on the $Na_2^+$ ions produced by PAI in a conventional Na MOT. They succeeded in determining the internal vibrational level distribution of $Na_2^+$ produced by the cold collision, the authors interpreted the distribution to infer the mechanism for this PAI process.

### 5.3.2  PAI and molecular hyperfine structure

Exploring the off-resonance detuning dependence of PAI, Wagshul *et al.* [417] discovered important effects of molecular ground-state hyperfine structure. With a MOT loaded from a Zeeman-slowed atomic beam, together with an extra probe laser, Wagshul *et al.* [417] alternated trapping and probing with a 50% duty cycle. Trap on/off periods varied from 2 to 100 μs, and ions were counted only during the probe period. In this scheme a large range of probe laser detuning could be explored without the complicating effects of the MOT fields (trap and repumper) and without compromising the trap performance. The total probe laser intensity ($\sim 3$ W cm$^{-2}$) was much greater than the trapping laser ($\sim 35$ mW cm$^{-2}$) but had a similar geometrical configuration and polarization to minimize perturbations on the MOT. Figure 5.9 shows the spectrum obtained by Wagshul *et al.* [417] in the region between the two hyperfine ground-state transitions, $F = 2 \rightarrow F' = 3$ and $F = 1 \rightarrow F' = 2$. A broad resonance structure rises sharply about halfway between the two ground-state hyperfine transitions and extends to the blue. The quadratic dependence of the amplitude of this feature with the probe laser intensity is consistent with a two-photon process. Observing variations in the peak intensity as the relative population between the two ground-state hyperfine levels were varied, Wagshul *et al.* [417] concluded that the observed spectrum arises from the collision of two ground-state atoms, each of which is in a different ground-state hyperfine level. The interpretation of Fig. 5.9 can be understood in the following way. Atoms in different hyperfine ground-states collide on the potential labeled $F = 1 + F = 2$. The

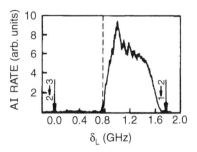

**Fig. 5.9.** Two-photon PAI spectrum arising from collisions on different ground-state hyperfine levels, from Wagshul *et al.* [417].

photon required for a two-step excitation to the doubly excited level falls halfway between the potentials labeled $1 + P$ and $2 + P$. Where the attractive $2 + P$ potential crosses this line, the transition becomes doubly resonant. The observed threshold for this process is within 10 MHz of the prediction. Light to the red of this frequency cannot provide enough energy to reach the $P + P$ potential starting from the $F = 1 + F = 2$ asymptote of the ground-state. Light to the blue of the threshold frequency follows the following route. First the optical field couples to the $2 + P$ curve at larger internuclear separation. The atoms then begin to accelerate together on the $2 + P$ potential until they reach a Condon point where the exciting light becomes resonant with a transition to the $P + P$ curve. This broad PAI feature shows substructure consistent with free–bound transitions to an intermediate attractive state. Wagshul *et al.* [417] carried out a model calculation of the PAI feature by applying their earlier two-step model (Julienne and Heather [201], Heather and Julienne [166]) and incorporating the ground-state hyperfine levels. The calculated structure resembles the experimental feature, but appears to be less congested; probably because the model neglects the multitudinous number of excited-state hyperfine levels.

### 5.3.3  Two-color PAI

The work of Wagshul *et al.* [417] brought to light the importance of ground-state hyperfine structure in early studies of PAI spectroscopy. They used a one-color probe to excite the two-step PAI process. Reference [24] investigated the same PAI processes, using two different laser frequencies, one of which could be varied independently, and starting with different initial states for the colliding atoms. Atomic sodium was cooled and trapped in a MOT loaded from the low-velocity tail of background vapor. The trapping frequency $\omega_1$ was red-detuned as usual about one line width from the cycling transition, $3s\,^2S_{1/2}(F = 2) \to 3p\,^2P_{3/2}(F' = 3)$; and the repumper $\omega_2$ was tuned close to $3s\,^2S_{1/2}(F = 1) \to 3p\,^2P_{3/2}(F' = 2)$. Depending on their relative intensity, these two frequencies could pump different relative populations into the two ground-state hyperfine levels $F = 1$ and $F = 2$. A channeltron detector measured the PAI ion signal as a function of the frequency of a third, tunable probe laser $\omega_p$. Figures 5.10(a) and (b) show the PAI profiles as $\omega_p$ scans to the blue of $\omega_1$ and $\omega_2$. In Fig. 5.10(a) the intensity of $\omega_1$ is stronger than $\omega_2$ by a factor of 3, and the inset above the line profiles shows the interpretation of the excitation routes. Both atomic hyperfine levels have significant population so $\omega_1$ and

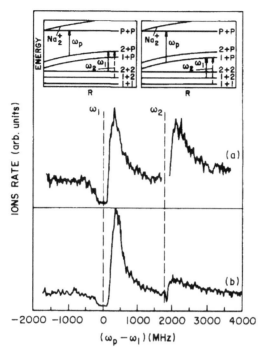

**Fig. 5.10.** Two-color photoassociative ionization. The inset shows MOT ($\omega_1$) and repumper ($\omega_2$) fixed frequencies and the sweeping probe ($\omega_p$) frequency, from [24].

$\omega_2$ can couple all three molecular entrance channels ($1 + 1$, $1 + 2$, $2 + 2$) to the $1 + P$ and $2 + P$ attractive excited-state levels. Then $\omega_p$ scanning to the blue couples these two singly excited levels to the doubly excited autoionizing $P + P$ curve. To test this interpretation Bagnato *et al.* [24] reversed the relative intensities in $\omega_1$ and $\omega_2$ so as to pump most of the atomic population to the $F = 2$ level. The effect is to depopulate the $1 + 1$ molecular entrance channel, and Fig. 5.10(b) shows that the amplitude of the PAI profile degrading to the blue of $\omega_2$ is greatly attenuated, which is consistent with this entrance channel depopulation. A common feature of all the measurements is the asymmetry in the PAI profile. As the probe laser scans to the blue, starting from the atomic transitions near $\omega_1$ and $\omega_2$, the PAI signal rises rapidly to a peak ~400 MHz detuning from the atomic limit, and then exhibits a slow decay over a range of about 2 GHz. The asymmetric shape of this two-color PAI spectrum was modeled by Gallagher [143] and arises from the action of four factors: (1) first excitation from the ground-state at the initial Condon point $R_i$, (2) spontaneous emission as the atoms approach on the first attractive excited state until they reach the second Condon point $R_2$, (3) second excitation from the first attractive state to the doubly excited state at $R_2$, (4) spontaneous emission as the atoms approach the autoionization zone on the doubly excited state. Once the atoms are within $R_{PAI}$, the zone where the quasimolecule can autoionize, the four excitation-decay factors must be multiplied by the autoionizing probability in order to calculate an overall ion production rate. Detuning of $\omega_p$ to the blue effectively increases the time between the first and second optical excitations. For short

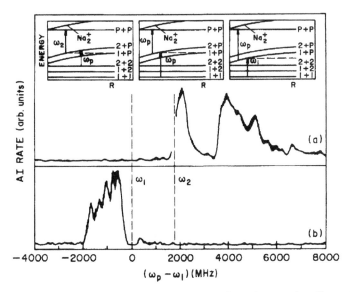

**Fig. 5.11.** Two-color photoassociative ionization. The insets above features describe excitation routes. Frequencies are labeled as in the previous figure from [24].

times (small blue detunings) the initially excited population does not have time to undergo appreciable radiative decay before excitation to the doubly excited level, yet the strong accelerating influence of the attractive $C_3/R^3$ potential shortens the travel time between $R_2$ and $R_{PAI}$ where the atoms enter the autoionizing region. These two factors account for the rapid rise in PAI rate at small blue detunings. As $\omega_p$ continues to scan to the blue, radiative depopulation of the first excited state starts to become significant and the overall PAI probability diminishes slowly. For atoms surviving to $R_2$ on the first excited state, the increased radial acceleration tends to minimize radiative decay on the doubly excited state once the population undergoes the second excitation at $R_2$. However, the probability of that excitation begins to diminish with increased acceleration because the atoms pass through the optical resonance around the Condon point $R_2$ with greater velocity. The overall effect is a slow diminution of the PAI production rate as $\omega_p$ scans further to the blue. Although the asymmetric shape calculated by Gallagher is in satisfactory accord with the measurements of Bagnato *et al.* [24] the absolute values of the rate constants differ by about an order of magnitude. Furthermore the position of the peak in the measured PAI profile is blue-shifted from Gallagher's model by about 200 MHz. These discrepancies could be accounted for by effects discussed previously in the chapter on trap loss, specifically (1) neglect of hyperfine structure and (2) semiclassical treatment at small detuning.

Reference [24] also investigated two-color PAI by prepumping all the atoms into either $F = 1$ or $F = 2$ and probing with only two frequencies: $\omega_p, \omega_1$ or $\omega_p, \omega_2$. Figures 5.11(a) and (b) shows the results with insets of the excitation routes. Panel (a) shows the spectrum with two principal features when only $F = 1$ is populated so that incoming scattering flux appears only on the $1 + 1$ asymptote. The onset of the right most features occurs at about twice the ground-state hyperfine splitting to the blue of $\omega_1$. The asymmetric shape degraded to the blue is characteristic of initial excitation to an attractive state at long range by fixed frequency $\omega_1$ followed by a second excitation to the doubly excited level by scanning frequency $\omega_p$. The right-hand inset above the feature shows the excitation pathway. It appears that the profile has structure which must come from bound states of the doubly excited manifold. This structure remains to be investigated thoroughly. The central peak, immediately to the blue of $\omega_2$, is quadratic in the probe laser intensity, suggesting a two-photon single-frequency process similar to that observed by Wagshul et al. [417]. The inset shows that $\omega_p$ carries population to some mixed state of $1 + P$ and $2 + P$, and a second photon takes the population to the doubly excited level. Panel (b) shows the spectrum obtained when only $F = 2$ is populated. Incoming scattering flux appears only on the $2 + 2$ asymptote, and the onset of the feature begins about 250 MHz to the red of $\omega_1$ and degrades to the red over a range of about 2 GHz. Structure is clearly visible and the inset above the feature shows that the scanning frequency $\omega_p$ excites vibration–rotation levels associated with a bound singly excited states while $\omega_2$ transfers population to the doubly excited level. Note that the left-and right-hand insets show how the fixed and scanning frequencies exchange roles. A PAI profile degrading to the blue means $\omega_p$ couples the strongly attractive singly excited state (or states) to the relatively flat doubly excited level above it; while a profile degrading to the red implies $\omega_p$ coupling from the strongly attractive manifold to the flat $2 + 2$ entrance channel below it. The fixed frequency $\omega_1$ or $\omega_2$ determines where on the absolute frequency scale these two-step excitation profiles appear. The red-degraded, structured profile reveals bound states of the singly excited manifold very close to the dissociation limit where molecular hyperfine coupling produces a complicated and congested spectrum. This structure was analyzed by Bagnato et al. [26] and found to be associated with a $0_g^-$ bound vibrational level progression, but extensive hyperfine mixing near dissociation render these symmetry labels ambiguous. Further probing to the red of the hyperfine region (Ratliff et al. [326]) reveals a much simpler, highly resolved spectrum, opening the very fruitful field of precision PAI spectroscopy, reviewed in the next section.

In follow-up studies, Jones et al. [196] have measured two-color spectra in which the first step populates a specific state in the $0_g^-$ or the $1_g$ Na$_2$ molecular electronic potentials. Then a second high-resolution light beam scans transitions either upward leading to autoionizing potentials near the $3P + 3P$ doubly excited asymptote or downward back to the manifold of potentials associated with the Na$_2$ ground electronic state. This study obtains spectroscopic constants for several levels of a doubly excited $1_u$ state, demonstrates the appearance of a new form of Condon internal diffraction from bound $\rightarrow$ free transitions just above the $3P_{3/2} + 3P_{3/2}$ dissociation limit, and discusses the use of photoassociation as a source of cold molecules. This work pioneers the high-resolution photoassociation spectroscopy of the doubly excited states of Na$_2$ and begins to unravel the similarities and differences between the mechanisms of associative ionization at high temperature and photoassociative ionization in the ultracold regime.

In a parallel theoretical study, Dulieu et al. [118] have shown that five molecular symmetries, $^1\Sigma_g^+, ^3\Sigma_u^+, ^1\Pi_u, ^3\Pi_u, ^1\Delta_g$, of the doubly excited states are likely to contribute to the molecular autoionization. They identify the peaks observed by Jones, et al. [196] and attributed to a doubly excited $1_u$ state (Hund's case (c) coupling), as belonging to the $^1\Pi_u$ member (Hund's case (a) coupling) of the five molecular symmetries. Exactly how

the short-range $^1\Pi_u$ joins to the long-range $1_u$ is still the subject of continuing investigation (Huynh *et al.* [185]. It appears, however, that the autoionization rate of this $^1\Pi_u$ state
precludes it from significantly contributing to the associative ionization process at thermal
energies.

## 5.4  Photoassociation spectroscopy in MOTs and FORTs

Although early cold alkali photoassociative ionization experiments explored different excitation and ionization pathways and confirmed the essential mechanisms of the process,
the full potential of photoassociation as a versatile and powerful spectroscopic technique
of unprecedented precision could not be realized until scanning probe fields could explore
beyond the hyperfine coupling zone of irregular and congested spectra. Two back-to-back
articles appearing in *Physical Review Letters* signaled a breakthrough in measuring photoassociation spectra well beyond this region. The NIST group continued to investigate
photoassociative ionization sodium, first with a strong probe laser and a conventional MOT
(Lett *et al.* [243], and then with a "dark spot" MOT which increased their ionization signal
about a hundredfold (Ratliff *et al.* [326]). In contrast the University of Texas group, led
by D. J. Heinzen, used a far-off-resonance optical trap (FORT) to capture rubidium atoms
(Miller *et al.* [279]) and to measure photoassociation spectra by measuring fluorescence
loss from the trap as a function of trap frequency (Miller *et al.* [280]. These two experiments opened a new domain of precision spectroscopy from which molecular potentials,
scattering dynamics, atomic lifetimes, and scattering lengths have all been determined with
an accuracy unattainable by conventional methods. A review of PAI has appeared (Lett
*et al.* [245]), lucidly explaining developments up to about mid-1995.

### 5.4.1  Sodium

The NIST group first reported a PAI spectrum (Lett *et al.* [243]) showing regular vibrational
progressions by scanning a probe laser to the red of the MOT cycling transition. The set-
up used the same conventional MOT and dedicated, retroreflected probe laser source that
Wagshul *et al.* [417] employed to study hyperfine effects. The PAI spectrum showed two
regular progressions, one strong series ranging from about 5 to 100 GHz to the red of
the MOT cycling transition and a second weaker series discernible from 15 to 40 GHz
red detuning. These regular features are the vibrational series corresponding to long-range
molecular potentials of the singly excited $Na_2$ molecule. The $C_3/R^3$ form of these long-range
potentials arises from the resonant dipole–dipole interaction between an excited $Na^*(3p)$
atom and a ground-state $Na(3s)$ atom. Four optically active excited molecular potentials
supporting vibrational levels and correlating to the $^2S_{1/2} + {}^2P_{3/2}$ asymptote are candidates
for the observed regular progressions. In Hund's case (c) coupling these states are labeled
$1_u, 0_g^-, 1_g$, and $0_u^+$. They will have different effective $C_3$ coefficients and their vibrational
progressions will have different hyperfine substructure. Both of these characteristics serve
to identify states to which the spectra belong.

LeRoy and Bernstein [242] and Stwalley [366] have shown that in this long-range regime the spacing between vibrational levels can be described analytically by,

$$\frac{dE(v)}{dv} = \hbar \left(\frac{2\pi}{\mu}\right) \cdot \frac{\Gamma(4/3)}{\Gamma(5/6)} \cdot \frac{3}{C_3^{1/3}} \cdot [D - E(v)]^{5/6} \tag{5.11}$$

which integrates to,

$$E_v - D = \left(\frac{a}{C_3^2}\right)(v - v_d)^6, \tag{5.12}$$

where $v$ is the vibrational quantum number counted from the top of the well and $v_d$ is the effective quantum number at the apparent dissociation. The number $v_d$ is a fraction between 0 and 1. The first bound vibrational level below dissociation is always labeled $v = 1$. The term $D$ denotes the dissociation energy and $a$ is a collection of constants including the reduced mass, $\mu$. The NIST group fit a plot of

$$(E_v - D)^{1/6} \text{ vs. } (v - v_d)$$

to the centroids of the peaks for the strong and weak series. From the slope they determined $C_3 = 10.37 \pm 0.07$ a.u. for the strong series and $C_3 = 10.8 \pm 0.7$ a.u. for the weak series. The calculated Hund's case (c) $C_3$ constants for the optically active excited states clearly narrowed the possible identification of the strong and weak seres to $1_g$ and $0_u^+$. The measured peak widths, due to hyperfine substructure, were $\sim 1$ GHz, consistent with calculations for $1_g$ but not for $0_u^+$. According to Williams and Julienne [446], the $0_u^+$ state should show about one-third of the hyperfine substructure of the $1_g$ state. Therefore, *both* the strong and weak series must belong to $1_g$. Since the apparent dissociation energy determined from Eq. 5.12 for the two series differs by about the ground-state hyperfine splitting, Lett *et al.* [243] concluded that the origin of the strong and weak progressions arose from $2 + 2$ collisions and $2 + 1$ collisions, respectively. However, one may ask why it is that the average measured $C_3$ lies near 10.6 a.u. with an uncertainty of about 0.7 while the best $C_3$ value for the $1_g$ state is 9.55 a.u., a difference of about eight. The reason is that the $1_g$ potential originates from diagonalizing a $3 \times 3$ potential matrix, and $C_3$ is only independent of $R$ in the region of internuclear separation where the binding energies of the states is much less than the atomic fine structure splitting, i.e. $C_3/R^3 \ll \Delta E_{FS}$. In this range of $R$ the $C_3$ term is *independent* of $R$. At shorter internuclear separation the diagonalized adiabatic states will fit to $C_3$ constants that depend on the range of $R$ over which the potential is sampled. Williams and Julienne [446] report that fitting vibrational eigenvalues obtained from adiabatic electronic-rotational-hyperfine molecular potentials to Eq. 5.12 yields a $C_3$ in agreement with the measured value of the strong series within experimental error *as long as the measured and calculated values span the same range in the progression*. The $C_3$ extracted from such fits can be used to obtain a precise measurement of the excited atom lifetime. Section 5.6 reviews precision measurement of excited state lifetimes.

Ratliff *et al.* [326] improved the NIST experiment by substituting a "dark spot" MOT (Ketterle *et al.* [215]) for the conventional MOT, increasing the ion signal by two orders of magnitude and permitting a probe scan over $172$ cm$^{-1}$ to the red of the trapping transition.

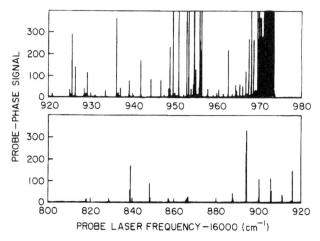

**Fig. 5.12.** Photoassociative ionization spectrum of Na collisions from the NIST dark-spot MOT, from Ratliff *et al.* [326].

Figure 5.12 shows the full range of this photoassociative ionization spectrum. These new results revealed the $1_g$, $0_u^+$, and $0_g^-$ states, identified by the energy range over which they were observed, vibrational and rotational spacing, and the extent of the hyperfine structure. Furthermore, alternately chopping the trap and probe lasers at 50 kHz while slowly scanning the probe, permitted two kinds of measurements: (1) trap loss spectra during the trap phase, as measured by ions produced by the trapping lasers, and (2) ionization production during the probe phase. Figure 5.13 shows how these two complementary spectra helped distinguish the various progressions. The long progression of the $1_g$ state shows clearly the change of coupling regime from Hund's case (c) to Hund's case (a) at closer range. Hund's case (a) labels the state $^1\Pi_g$ and the vibrational lines split into rotational progressions. The molecular transition giving rise to this progression is $^3\Sigma_u \rightarrow 1_g$ (or $^1\Pi_g$ at close range); but the triplet character diminishes as the coupling scheme transforms from Hund's case (c) to (a), and the natural line width diminishes to less than 0.1 MHz. Therefore at closer range (detunings $\geq 50$ cm$^{-1}$), the spectral resolution is limited only by the probe laser width ($\sim 2$ MHz) permitting analysis of the line profiles of the resolved rotational progressions.

The individual rotational line profiles show structure and asymmetries due to the finite temperature of the MOT, residual hyperfine splitting and quantum threshold effects. Figure 5.14 shows the measured rotational line shapes obtained by Ratliff *et al.* [326] for the $v = 48$ vibrational level of the $1_g$ state. Evidence of hyperfine contributions can be seen in the incompletely resolved substructure of $J = 1$ and $J = 2$ profiles. An analysis of these rotational line shapes have been carried out by Napolitano *et al.* [302], the results of which are shown in Fig. 5.15. The intensity of the photoassociation spectrum is proportional to the thermally averaged rate coefficient. The quantum scattering expression for the rate coefficient in terms of the scattering matrix, $S$, and the scattering partial waves, $l$,

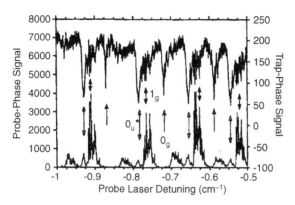

**Fig. 5.13.** The top spectrum shows trap-loss dips as the probe laser scans through molecular rotational progressions. The bottom ion spectrum shows complementary photoassociative ionization signals, from Ratliff *et al.* [326].

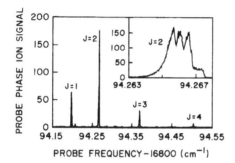

**Fig. 5.14.** A high-resolution PAI spectrum of rotational levels associated with $v = 48$ of the long-range $1_g$ state of $Na_2$. Note the residual hyperfine splitting in the $J = 2$ inset, from Ratliff *et al.* [326].

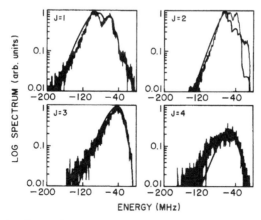

**Fig. 5.15.** High-resolution line profile measurements and theory on rotational progression associated with $v = 48$ of the long-range excited $1_g$ state of $Na_2$, from Napolitano *et al.* [302].

is written as

$$K_p(T, \omega) = \frac{k_B}{h Q_T} \sum_{l=0}^{\infty} (2l + 1) \int_0^{\infty} |S_p(E, l, \omega)|^2 \, e^{-E/k_B T} \frac{dE}{k_B T}, \tag{5.13}$$

where $S_p(E, l, \omega)$ is the $S$-matrix element for product channel p, $Q_T = (\frac{2\pi \mu k_B T}{h^2})^{3/2}$ is the translational partition function, $k_B$ is Boltzmann's constant, $E = \hbar k^2/2\mu$ is the asymptotic kinetic energy of the colliding ground-state $^2S$ atoms, and the other terms in the expression have their usual meanings. At relatively large detunings where the vibration–rotation lines are isolated from each other, Napolitano *et al.* [302] used quantum scattering calculations with a complex potential to show that the form of $S_p(E, l, \omega)$ around a scattering resonance can be modeled as

$$|S_p(E, l, \omega)|^2 = \frac{\Gamma_p \Gamma_s(E, l)}{(E - \Delta_b)^2 + (\Gamma/2)^2}, \tag{5.14}$$

where $\Delta_b$ is the probe laser detuning from the bound rovibronic state of the excited level of the free–bound transition, $\Gamma_p$ is the decay width to the collision product (whatever that decay process might be) and $\Gamma_s(E, l)$ is the stimulated rate connecting the ground and excited levels. The $\Gamma$ term in the denominator is the sum of $\Gamma_p$ and $\Gamma_s$, plus any other decay processes that affect the level in question. Since $\Gamma_s = 2\pi P_{ge}(E)$, where $P_{ge}(E)$ is defined by Eq. 2.53, $\Gamma_s$ exhibits Wigner-law threshold behavior due to the the quantum properties of the ground-state wavefunction as described in Section 2.2. This threshold behavior has the form of $\Gamma_s \approx A_l E^{(2l+1)/2}$ as $E \to 0$. Three partial waves, s, p, d contribute to the scattering amplitude of the $J$-levels of Fig. 5.15; and their influence can most easily be seen on the blue (right-hand) slope of each profile. Note especially in the $J = 2$ panel of Fig. 5.15 the very sharp onset to the profile, illustrating the influence of the Wigner threshold law for the s-wave contribution (close to infinite slope as $E \to 0$). On the red side of the profile, the intensity falls off as $e^{-\Delta_b/k_B T}$ and at first thought one might expect to extract a collision temperature from the slope of a semilog plot. However, the s, p, d contributions to the line shapes all yield somewhat different effective temperatures, again due to the influence of the Wigner threshold law operating on $\Gamma_s$. In order to extract a meaningful temperature from the experimental line shape therefore the fractional partial wave contributions to the profile must be known. It is also worth noting that the particular admixture of partial waves at a particular temperature will shift the peak positions of the lines, and the line "position" must be found by modeling the onset on the blue side of the profile (Jones *et al.* [194]).

Later, a combination of high-resolution photoassociative spectroscopy of cold Na together with conventional molecular spectroscopy using atomic beams and a heat pipe, permitted Jones *et al.* [195] to perform a direct measurement of the ground-state dissociation energy ($D_0$) of Na$_2$ with unprecedented precision. Combining three spectroscopic measurements, Jones *et al.* [195] obtained $D_0$ for the X $^1\Sigma_g^+$ ground-state as $D_0 = 5942.6880(49) \, \text{cm}^{-1}$, making this the most accurately known dissociation energy of any chemically bound neutral molecule. This result demonstrates how cold photoassociation spectroscopy can comple-ment conventional spectroscopy with a considerable increase in resolution.

Additional two-color experiments have revealed other interesting details about the PAI process. Molenaar *et al.* [290] have reported a two-color PAI spectrum in an Na MOT in which they determine high-lying vibrational levels of the $0_g^-$ and observe a predissociation of this state at a curve-crossing interaction between the $0_g^-$ and the ground-state $^2S_{1/2}(F = 1)$ hyperfine level. Similar spectra near the $0_g^-$ dissociation have been reported by Bagnato *et al.* [26]. In a follow-up paper to [24] Marcassa *et al.* [256] studied the overall profile of the two-color photoassociative ionization spectra. They showed that a local-equilibrium, semiclassical theory can explain the main features of the red-detuning profile.

### 5.4.2  Rubidium

Although the atom trap most commonly used for collision studies has been the MOT, Miller *et al.* [279] pioneered the use of the far-off-resonance trap to study photoassociation exper-iments in Rb. The advantages of the FORT over the MOT are: (1) relative simplicity (no magnetic fields); (2) very low excited-state population and therefore little diffusional heat-ing; (3) high density ($10^{12}$ cm$^{-3}$); and (4) a well-defined polarization axis along which atoms can be aligned or oriented. The disadvantage is the relatively low number of atoms ($\sim 10^4$) trapped in the small volume ($\sim 5 \times 10^{-9}$ cm$^3$). In the experiment of Miller *et al.* [280], the FORT was loaded from a MOT by spatially superposing the focused FORT laser on the MOT and alternately chopping the two kinds of traps. At the end of the load sequence fluorescence detection probes the number of atoms in the FORT. In the first version of the experiment, the FORT laser itself was swept in frequency from 50 to 980 cm$^{-1}$ to the red of the atomic resonance. As the laser scanned through the bound–state resonances, the free–bound association peaks were detected by measuring the resulting loss of atoms from the FORT. Figure 5.16 shows the results of the trap-loss spectra. The increase of the background intensity as the dissociation limit is approached represents an increase in the Franck–Condon overlap and is also a consequence of the increase in the atomic density in the trap. The solid and dashed vertical lines on the top of the spectra of Fig. 5.16 identify two vibrational series. Comparison of these series with energy eigenvalues of model potentials by Movre and Pichler [297], Krauss and Stevens [231], and Bussery and Aubert-Frecon [65, 64] identify them as the $1_g$ and $0_g^-$ states. A noticeable feature in the spectrum ob-tained by Miller *et al.* [280] is the slow oscillation in the intensity of the lines. Thorsheim *et al.* [385] predicted these "Condon modulations". They arise from the Franck–Condon over-lap of the ground- and excited-state wave functions, and the oscillations reflect the nodes and antinodes of the scattering wavefunction on the entrance ground-state channel. This is a consequence of the refection principle in Eq. 2.57. Changing the excitation frequency changes the Condon point $R_C$ so that the spectrum tracks the shape of the ground-state wavefunction.

A second version of this experiment [77, 78] separated the trapping and scanning functions with two different lasers: one laser at fixed frequency produced the FORT while a probe laser

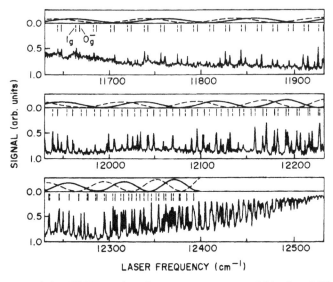

**Fig. 5.16.** Photoassociation FORT trap-loss fluorescence spectrum of $Rb_2$, from Miller *et al.* [280]

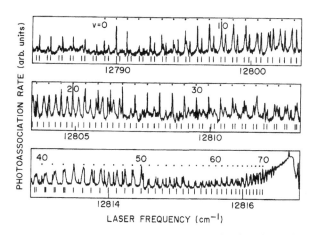

**Fig. 5.17.** Cline *et al.* [77, 78] separated the trapping and scanning functions with two different lasers: one laser at fixed frequency produced the FORT while a probe laser scanned through the resonances.

scanned through the resonances. The two lasers, probe and FORT, were alternately chopped to avoid spectral distortion from AC Stark shifts and power broadening. Miller *et al.* [280] restricted themselves to a spectral region from 50 cm$^{-1}$ to 980 cm$^{-1}$ below the $^2S_{1/2} +^2 P_{1/2}$ dissociation limit of $Rb_2$. The two-color experiment [77, 78] enabled a greatly increased spectral range and the identification of the $0_g^-$ "pure-long-range" state, the $1_g$, and the $0_u^+$. The high-resolution photoassociation spectrum is shown in Fig. 5.17. From this high-resolution

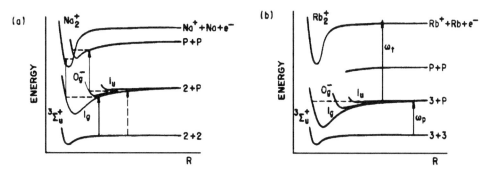

**Fig. 5.18.** PAI in sodium and rubidium collisions: (a) potential diagram of $Na_2$ and $Na_2^+$ showing doubly excited asymptote above $Na_2^+$ ionization threshold; (b) potential diagram of $Rb_2$ and $Rb_2^+$ showing the doubly excited asymptote below the $Rb_2^+$ threshold.

spectra the Texas group [77, 78] determined the $C_3$ coefficient for two molecular excited states of $Rb_2$. The values are: $C_3(0_u^+) = (14.64 \pm 0.7)$ a.u. and $C_3(1_g) = (14.29 \pm 0.7)$ a.u. The spectra also show that the $0_u^+$ lines are broadened by predissociation to the $5\,^2S_{1/2} + 5\,^2P_{1/2}$ asymptote from spin–orbit mixing of the $0_u^+$ components of $A\,^1\Sigma_u^+$ and $b\,^3\Pi_u$ at short range. This broadening is evidence of the FCC trap loss process discussed in Chapter 4. Recently the Texas group, in collaboration with the Eindhoven theory group, extended their experiments to a two-color photoassociation spectroscopy in which the second color couples population from the initially excited $0_g^-$ to the last 12 vibrational levels of the $^{85}Rb_2$ ground-state (Tsai *et al.* [401]. This high-precision spectroscopic data permits accurate determination of interaction parameters and collision properties of Rb ultracold collisions including the scattering length and the position of Feshbach resonances (see Section 7.2), both of which are very important for BEC in ultracold Rb.

As an alternative to trap-loss spectroscopy Leonhardt and Weiner [241] investigated $Rb_2$ photoassociation spectra by direct photoassociative ionization. Figure 5.18(b) illustrates the main steps in this experiment. Atoms held in a vapor-cell-loaded MOT are illuminated with two light beams: a probe beam $\omega_p$ scans through the resonances of the singly excited $Rb_2$ long-range potential while a second, transfer laser at fixed frequency $\omega_t$ directly pho-toionizes the bound excited molecule to high-lying vibrational states of $Rb_2^+$. This direct two-color photoassociative ionization process does not compete with an open autoionizing channel, as in sodium (Fig. 5.18a), because the manifold of doubly excited states $P + P$ lies below the minimum in the dimer ion potential. Sweeping the probe laser to the red of the atomic resonance and counting the ions, Leonhardt and Weiner [241] produced the pho-toassociative spectrum shown in Fig. 5.19 where the FORT trap-loss spectrum obtained by Cline *et al.* [77, 78] has been added for comparison. The direct $Rb_2$ PAI spectrum appears to show only one molecular state progression. In fact, comparison with the FORT experiment shows that the $0_g^-$ progression is absent in the photoassociative ionization spectrum. The measured series is assigned to the $1_g$ state and the determination of the potential constant resulted in $C_3(1_g) = (14.43 \pm 0.23)$ a.u. in good agreement with the value determined by

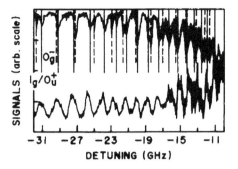

**Fig. 5.19.** Direct photoassociative ionization of $^{85}$Rb collisions in a MOT. The upper trace is the trap loss spectrum from Texas FORT, and the lower trace is the $Rb_2^+$ ion signal from the direct PAI, from Leonhardt and Weiner [241].

[77, 78]. Direct PAI should be widely applicable to many trappable species, but requires the densities of a dark spot MOT or a FORT to compete effectively with fluorescence trap loss spectroscopy.

### 5.4.3  Lithium

The Rice University group, lead by R. G. Hulet, has performed series of experiments in an atomic-beam-loaded MOT of $^6$Li and $^7$Li. Photoassociation trap loss fluorescence spectroscopy has been carried out on both isotopes and the resulting data has been used to determine $C_3$ coefficients, atomic lifetimes, and scattering lengths. The Rice group reported the basic spectroscopy, experimental technique, and analysis late in 1995 [5]. Earlier in the year the same group determined a precise atomic radiative lifetime for the Li(2p) level using the same photoassociation spectral data (McAlexander *et al.* [266]). A sample of the $^6$Li$_2$ spectrum reported by McAlexander *et al.* [266] is shown in Fig. 5.20. Indicated in the figure are the vibronic series for the $1\,^1\Sigma_u^+$ and $1\,^3\Sigma_g^+$ states which converge on the $2s\,^2S_{1/2} + 2p\,^2P_{1/2}$ asymptotic limit. Three rotational lines ($N = 0$, 1 and 2) are resolvable in the $1\,^3\Sigma_g^+$ series as shown in the inset of Fig. 5.20. The $N = 0$ line is used for the series analysis to construct part of the $1\,^3\Sigma_g^+$ interaction potential. The basic approach is to fit together a molecular potential at short, medium, and long range from various sources of calculated potentials and spectral data. At long range beyond 20 $a_0$ the potential is approximated analytically by the form

$$V \simeq -\frac{C_3}{R^3} - \frac{C_6}{R^6} - \frac{C_8}{R^8}, \tag{5.15}$$

where $C_3$ is due to the dominant resonant dipole interaction, and the other two dispersion terms are quite small compared to the dipole term. At very short range, less than 4.7 $a_0$, and then from 7.8 to 20 $a_0$ the potential uses *ab initio* calculations. In the region between 4.7 and 7.8 $a_0$ conventional spectroscopy yields an accurate Rydberg–Kline–Reese (RKR) potential. The different potential segments are then pieced together with a cubic spline and the the spectral positions are calculated while $C_3$ is varied as a free parameter until the best

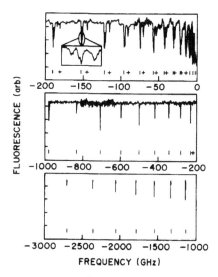

**Fig. 5.20.** Photoassociation spectrum from a $^6$Li MOT. The inset shows rotational substructure on vibrational lines, from McAlexander *et al.* [266].

fit is obtained. McAlexander *et al.* [266] finally report $C_3(1\,^3\Sigma_g^+) = 11.048 \pm 0.066$ a.u. This $C_3$ is then used to calculate a lifetime for the Li(2p) level. Deriving lifetimes from photoassociation spectra will be discussed in Section 5.6. In more recent work the same group [4] reported spectra of hyperfine resolved vibrational levels of the $^1\Sigma_u^+$ and $^3\Sigma_g^+$ states of both lithium isotopes, obtained via photoassociation. A simple first-order perturbation theory analysis accounts for the frequency splittings and relative transition strengths of the observed hyperfine features.

### 5.4.4 Potassium

The research groups of W. C. Stwalley and P. L. Gould at the University of Connecticut have joined forces to mount a comprehensive study of the photoassociation spectroscopy in ultracold $^{39}$K collisions. The implementation of a MOT in potassium presents an unusual situation among the alkalis. All the excited-state hyperfine levels are spaced within 34 MHz, making it impossible to identify well-defined cycling and repumping transitions to cool and maintain the MOT. Nevertheless, following Williamson and Walker [448], the Connecticut group was able to realize a working potassium MOT by essentially treating the whole group of excited hyperfine levels as a single "line" and tuning the cooling laser 40 MHz to the red of the $^{39}$K $4s\,^2S_{1/2}(F = 2) \to 4p\,^2P_{3/2}(F' = 3)$ transition. The "repumping" beam was set 462 MHz to the red of the same transition, equal to the splitting between the ground-state hyperfine levels. A potassium MOT requires relatively high intensity: 90 mW cm$^{-2}$ for the "cooling" laser and 40 mW cm$^{-2}$ for the "repumper," although both beams undoubtedly fulfill both cooling and repumping functions. Wang *et al.* [427] report a MOT performance

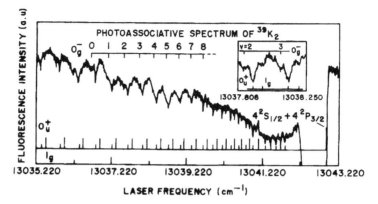

**Fig. 5.21.** Trap-loss fluorescence spectrum from photoassociation of $^{39}$ K, from Wang *et al.* [427].

of $\sim 2 \times 10^7$ captured atoms, a density of $\sim 3 \times 10^{10}$ cm$^{-3}$ at a temperature of $\sim 500$ μK. Insertion of a dark-spot disk (Ketterle *et al.* [215]) in the horizontal components of the "repumper" beam enhanced the photoassociation rate by about one order of magnitude. With this potassium MOT, the Connecticut group carried out cold photoassociation spectroscopy by measuring trap-loss fluorescence as an intense (1–2 W cm$^{-2}$) probe laser scanned about 7 cm$^{-1}$ to the red of the $S_{1/2}(F = 2) \rightarrow P_{3/2}(F' = 3)$ reference. Figure 5.21 shows an example of the regular progressions observed. In their initial report (Wang *et al.* [427]) identified the $0_g^-$, $0_u^+$, and $1_g$ states associated with the $S_{1/2} + P_{3/2}$ asymptote. In a follow-up study on the "pure long-range" states, Wang *et al.* [426] determined the molecular constants of $0_g^-$ and the $C_3$ constants for the $0_u^+$ and $1_g$ states. They also pointed out the utility of $0_g^-$ as an intermediate "window" state to access even higher-lying states K$_2$ by optical–optical double resonance. Wang *et al.* [433] have observed optical–optical double resonance PAI spectroscopy near the $^{39}$K(4s) + $^{39}$K(4d, 5d, 6d, 7s) asymptotes, using the $v' = 0$ level of the $0_g^-$ as the intermediate "window." The Connecticut team has also reported a comprehensive study of the long-range interactions in potassium binary atom collisions (Wang *et al.* [429]. They have observed the six long-range states of $0_u^+$, $1_g$, and $0_g^-$ symmetry, three dissociating to the 4s $^2$S$_{1/2}$ + 4p $^2$P$_{3/2}$ asymptote and three to the 4s $^2$S$_{1/2}$ + 4p $^2$P$_{1/2}$ asymptote. Furthermore, their analysis has determined the $C_3^{\Pi}$ and $C_3^{\Sigma}$ constants from the $0_g^-$ "pure long-range" state, from which they extract a precise measure of the K(4p) radiative lifetime, as well as the $C_6^{\Pi}$ and $C_6^{\Sigma}$ constants. More recently, they have identified for the first time in an alkali dimer spectrum the pure long-range $1_u$ state (Wang *et al.* [433]). Finally, Wang *et al.* [429] have reported direct observation of predissociation due to fine-structure interaction in the photoassociation spectra of K$_2$.

## 5.5  Photoassociative ionization in atom beams

As discussed earlier in Section 3.2, collisions in an atom beam can reveal polarization, orientation, and alignment properties that are averaged out in "cell" experiments using MOTs.

Until the advent of cold, bright beams, no PAI spectra exhibiting alignment and orientation information had been measured in intra-beam collisions due to low beam density and relatively high divergence. However, DeGraffenreid et al. [106] have developed a dense, optically brightened beam specifically to meet the requirements of collision studies. In addition to low angular divergence (high brightness) combined with low velocity dispersion (high brilliance) the atom beam density of $(1 \pm 0.5) \times 10^{10}$ cm$^{-3}$ is comparable to that obtained in conventional MOTs. This arrangement produces an intra-beam PAI collision rate sufficient for high-resolution PAI spectroscopy.

In general, even if an atomic beam is very bright and brilliant, collisions within it will be subject to spatial averaging if the final scattering channel is coupled to the wide acceptance angle (in the molecular coordinate frame) of the reactant entrance channel. However, when the entrance channel angle is very narrow, the molecular collision axis and the atom beam laboratory axis will superpose; and fortunately PAI falls into this category. As discussed in Section 3.2.3 and shown in the Newton diagram (Fig. 3.6), the range of possible molecular collision axes in PAI is restricted to a narrow acceptance angle determined by the long-range–short-range two-step mechanism expressed by Eq. 5.4 and illustrated in Fig. 5.7. Therefore aligning the polarization axis of the laser light parallel or perpendicular to the bright atom beam axis is tantamount to aligning it parallel or perpendicular to the molecular collision axis.

Ramirez-Serrano et al. [324] used a cold, bright sodium beam to study PAI in intra-beam collisions at a collision temperature of $\simeq 4$ mK. They recorded the one-color PAI rate as a function of red detuning from 0 to 45 GHz and as a function of the laser polarization alignment, parallel or perpendicular to the atom beam. Figure 5.22 shows the sensitivity to polarization and a comparison with a typical MOT spectrum in a restricted range of detuning, within the first 5 GHz to the red of the atomic transition, $3\,^2S_{1/2}(F = 2) \rightarrow 3\,^2P_{3/2}(F = 3)$. The comparison in Fig. 5.22 illustrates three significant differences between MOT and beam spectroscopy. The beam measurements evidently show the expected marked polarization dependence in the peak amplitudes, but they also show a slightly improved spectral resolution over the MOT results, and new peaks within a detuning range from 0 to 750 MHz where the photoassociation laser beam destabilizes a MOT.

In the first 5 GHz detuning, the two-step PAI process consists of an initial excitation to a "pure long-range" state [368] of $0_g^-$ symmetry, followed by a second step which transfers population to the doubly excited state of either $1_u$ or $0_u^-$ symmetry [13, 14]. The doubly excited state subsequently autoionizes at short range. Since the first excitation step must pass through a $0_g^-$ state, dipole selection rules dictate that the starting ground molecular state must be of $1_u$ or $0_u^-$ symmetry, corresponding to components of the familiar Na$_2\,^3\Sigma_u$ lowest triplet ("ground") state in Hund's case (b) notation. Electric dipole transitions in which the quantum number for the total electronic angular momentum projection along the quantization axis is unchanged or differs by one unit, $\Delta\Omega = 0\ (\pm 1)$, are called parallel (perpendicular) transitions because the molecular states are coupled by the dipole transition moment parallel (perpendicular) to the quantization axis. Hence, PA light with parallel (perpendicular) polarization only couples the $0_u^-$ ($1_u$) ground-states to the intermediate $0_g^-$ state. Starting from the common $0_g^-$ intermediate state, parallel (perpendicular)

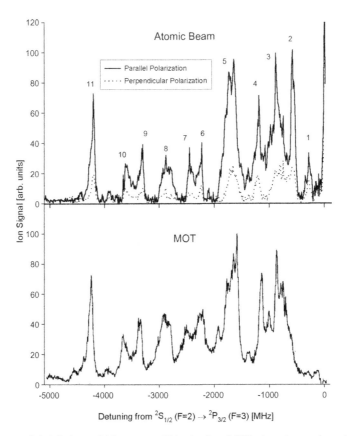

**Fig. 5.22.** Top panel: bright beam PAI spectra within the first 5 GHz detuning to the red of resonance. The solid trace denotes parallel polarization and the dotted trace denotes perpendicular polarization. Peak numbered labels are for reference only. Bottom panel: PAI spectra obtained from a conventional MOT, from Fig. 5 of [13].

polarized light populates the doubly excited state $0_u^-$ ($1_u$) symmetry. Thus, as summarized in Fig. 5.23, light polarized parallel or perpendicular to the atom beam axis populates selectively the doubly excited state of either symmetry from the ground-state. These excitation routes in Fig. 5.23 show that the second step should populate *either* the $0_u^-$ state (parallel polarization) *or* the $1_u$ state perpendicular polarization. Therefore, we would expect those peaks in Fig. 5.22 associated with $0_g^- \rightarrow 0_u^-$ to appear only with parallel polarization and the remaining peaks, associated with $0_g^- \rightarrow 1_u$, to appear only with perpendicular polarization. In fact, very nearly the same set of peaks appears with excitation by both polarizations, but with quite different intensities. An analysis of the intensity ratios for the two polarizations indicates that that the two doubly excited states may not be well characterized by the symmetry labels $0_u^-$ and $1_u$ and that the actual states may be about equal mixtures of these two. The symmetry labels had been assigned from a spectroscopic analysis of the rotational branches of the rovibrational levels in the PAI spectra [14]. The

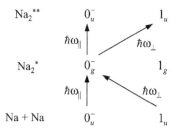

**Fig. 5.23.** Schematic diagram showing how two successive parallel or perpendicular transitions, passing through the $0_g^-$ intermediate state, selectively populate the $0_u^-$ or $1_u$ doubly excited states in the region of 0–5 GHz detuning.

polarization data appears to be at variance with this analysis, and the problem remains unresolved.

## 5.6 Atomic lifetimes from photoassociation spectroscopy

Because cold atoms collide at temperatures corresponding to kinetic energy distributions of the order of or less than the spontaneous emission line width of their cooling transition, they are particularly sensitive to binary interactions when the partners are far apart. The dominant long-range interaction between identical atoms, one resonantly excited and the other in the ground-state, is the dipole–dipole potential discussed many years ago by King and van Vleck [219]. They showed that this interaction potential between atom A in excited state P and atom B in ground-state S is given by,

$$V(R) = \pm \frac{1}{R^3}(\langle d_x \rangle^2 + \langle d_y \rangle^2 - 2\langle d_z \rangle^2), \tag{5.16}$$

where $\langle d_q \rangle$, $q = x, y, z$ are the transition dipole matrix elements between the two states. Normally the $z$ axis is taken as the line joining the two atom centers, and $C_3$ constants are then defined in terms of the transition dipoles as a "transverse" component,

$$C_3^{\Pi} = \langle d_x \rangle^2 = \langle d_y \rangle^2 \tag{5.17}$$

and an "axial" component,

$$C_3^{\Sigma} = 2\langle d_z \rangle^2 = 2C_3^{\Pi}. \tag{5.18}$$

The transition moments themselves are related to the atomic excited-state lifetime $\tau$ through the average rate of spontaneous photon emission $\Gamma$, written in MKS units as

$$\Gamma = \tau^{-1} = \frac{16\pi^3 v^3}{3\epsilon_0 h c^3} \cdot \langle d_q \rangle^2.$$

With $C_3^{\Pi}$ also written in MKS units,

$$C_3^{\Pi} = \frac{1}{4\pi \epsilon_0} \cdot \langle d_q \rangle^2 \tag{5.19}$$

the atomic lifetime can be written in terms of $C_3$ as

$$\tau = \frac{3\hbar}{4C_3^\Pi} \left(\frac{\lambda}{2\pi}\right)^3 . \tag{5.20}$$

Therefore, a precise determination of $C_3^\Pi$ from cold photoassociation spectroscopy will determine a precise measure of the atomic excited-state lifetime.

For most molecular states that exhibit strong chemical binding, however, the long-range, dipole–dipole interaction is only one of several $R$-dependent interaction terms in the Hamiltonian. At close range the electron clouds from each atom interpenetrate and give rise to electron–nuclear attraction, electron–electron repulsion terms, as well as attractive quantum mechanical electron exchange terms which decrease exponentially with increasing $R$. In this near zone the electron orbital angular momentum, with a molecular axis projection denoted by $\Lambda$, strongly couples to the internuclear axis, and the electron spin $S$, orbital angular momentum, and axis rotation angular momenta are coupled in a scheme called Hund's case (a), with states designated by the notation $^S\Lambda$. At greater internuclear separation, beyond the exchange interaction, the interatomic interaction is described by resonant dipole–dipole and dispersion terms varying with inverse powers of $R$, $V(R) \sim \pm C_3/R^3 - C_6/R^6 - C_8/R^8$. At sufficiently large distances, where the energy separations between the different Hund's case (a) states become small compared to the atomic fine-structure splitting, electron orbital and spin angular momentum couple strongly together to form a total electronic angular momentum, whose projection on the molecular axis is designated $\Omega$. The total electronic angular momentum couples with nuclear rotation to form a total angular momentum called Hund's case (c), with states labeled by $\Omega$. This scheme is used to label the potentials in Fig. 5.2.

In order to extract $C_3$ from observed spectral line positions, it is necessary to understand all the terms in the molecular Hamiltonian, which have contributions from the interatomic interactions, molecular rotation, and fine and hyperfine structure. Fortunately, these can all be taken into account with great precision and detail in the case of photoassociation, where the spectra are due to single molecular states. Although a complete quantitative theory of the rotating molecule with hyperfine structure can be developed (Tiesinga *et al.* [393]) it is much easier to use simpler models, which are also highly quantitative (Jones *et al.* [194]). The Hamiltonian of Movre and Pichler provides such a model. It includes the interatomic and fine-structure interactions (although rotation and hyperfine terms could be readily added). Only states with the same projection $\Omega$ of total electronic angular momentum on the interatomic axis can couple to one another ($\Omega = \Lambda + \Sigma$, where $\Lambda$ and $\Sigma$ are the respective projections of orbital and spin angular momenta on the axis). At long range the off-diagonal interatomic interaction due to the resonant dipole coupling, proportional to $C_3/R^3$, mixes different states of the same $\Omega$. Diagonalizing the Movre–Pichler Hamiltonian gives the adiabatic molecular potential curves shown in Fig. 5.2. Therefore each individual adiabatic curve, such as $1_g$ or $0_g^-$, has its own characteristic $C_3$ coefficient at long range.

A good example of the Movre–Pichler analysis is provided by the two pure long-range $0_g^-$ states dissociating to S + P. These result from a spin–orbit avoided crossing between two Hund's case (a) basis states: a repulsive $^3\Pi_{0g}$ potential, which goes as $+C_3/R^3$, and an attractive $^3\Sigma_{0g}$ potential, which goes as $-2C_3/R^3$. The two adiabatic $0_g^-$ potentials are found by diagonalizing the potential matrix (Jones *et al.* [194]):

$$
V_{MP} = \begin{array}{cc} \qquad\qquad \Pi \qquad\qquad\qquad \Sigma \\ \begin{pmatrix} \dfrac{C_3}{R^3} - \dfrac{2\Delta_{FS}}{3} & \dfrac{\sqrt{2}\Delta_{FS}}{3} \\[3mm] \dfrac{\sqrt{2}\Delta_{FS}}{3} & -\dfrac{2C_3}{R^3} - \dfrac{\Delta_{FS}}{3} \end{pmatrix} \begin{array}{c} \Pi \\[3mm] \Sigma \end{array} \end{array}
\tag{5.21}
$$

where $\Delta_{FS}$ is the atomic spin–orbit splitting ($\Delta_{FS} = 515.520$ GHz for Na) and the zero of energy is the $^2S_{1/2} + {}^2P_{3/2}$ asymptote. Diagonalizing this matrix gives the two $0_g^-$ states in Fig. 5.2, the upper going to $^2P_{3/2} + {}^2S_{1/2}$ and the lower to $^2P_{1/2} + {}^2S_{1/2}$. The upper $0_g^-$ potential has a shallow well and is shown in Fig. 4.2. Although both curves are characterized by only a single atomic transition dipole matrix element, the diagonalization is $R$-dependent, and the effective $C_3$ value associated with a particular adiabatic curve is $R$-dependent, changing from a long-range value characteristic of Hund's case (c) coupling to a short-range value characteristic of Hund's case (a) coupling. For example, the attractive Hund's case (c) $1_g$ state from $^2P_{3/2} + {}^2S_{1/2}$ changes to a Hund's case (a) $^1\Pi_g$ state at short range. We have already seen in Chapter 5 how the $C_3$ value found from fitting spectra for this state changes with the range of levels fitted. At short range the potentials are also affected by other terms in the electronic Hamiltonian. The van der Waals dispersion terms, $C_6/R^6$, first begin to play a role as $R$ decreases, then the exchange-overlap interaction due to overlapping atomic orbitals on the two atoms, and finally the strong chemical bonding at very small $R$. The effects of all of these terms must be taken into account, with the specific details depending on the particular state being analyzed. The $0_g^-$ state is especially useful in this regard, since it depends mainly on $C_3$ and to a lesser extent on the dispersion energy.

The Rice group used the lower $^3\Sigma_g^+$ state of $Li_2$ (McAlexander *et al.* [266]). Since this state is chemically bound with a large binding energy, the potential must be very accurately known over a wide range of $R$ in order to use $C_3$ as a free–variable fitting parameter to minimize differences between a constructed model potential and the measured data. The accuracy of the lifetime determination was limited to 0.6% by the use of *ab initio* calculated potentials over a range of $R$ (from 7.8 to 20 $a_0$) where spectroscopic data was unavailable. In an update to this experiment (McAlexander *et al.* [267]), the Rice group was able to shrink the error bars even more by applying essentially the same approach but using an experimentally derived potential over the range of $R$ where they had previously relied on calculations; and, by analyzing the hyperfine spectrum, as reported in [4], they reduced the uncertainty in the vibrational energies. The reported uncertainty in the Li(2P) lifetime from these measurements is 0.03%.

The "pure" long-range states, first proposed by Stwalley *et al.* [368], support bound vibrational levels with turning points entirely outside the region where exchange potential makes significant contributions. The molecular potential therefore includes only the power series terms with atomic coefficients $C_3, C_6, C_8, \cdots$ and the constant atomic spin–orbit term. Transition moments and the atomic lifetimes can then be calculated from Eqs. 5.19 and 5.20. The NIST (Jones *et al.* [194]) and Connecticut (Wang *et al.* [426]) groups have followed

**Table 5.1.** Atomic lifetimes and transition moments.

| Atom | Lifetime, $\tau$ (ns) | $d^2$ (a.u.) | |
|---|---|---|---|
| $^7\text{Li}^a$ $(2\text{P}_{1/2})$ | $27.102\ (2)^b\ (7)^c$ | $11.0028$ | $(8)^b(28)^c$ |
| $\text{Na}^d$ $(3\text{P}_{3/2})$ | $16.230\ (16)^b$ | $6.219$ | $(6)^b$ |
| $^{39}\text{K}^e$ $(4\text{P}_{3/2})$ | $26.34\ (5)^b$ | $8.445$ | $(14)^b$ |
| $\text{Rb}^f$ $(5\text{P}_{3/2})$ | $26.23\ (6)^b$ | $8.938$ | $(20)^b$ |
| $\text{Cs}^g$ $(6\text{P}_{1/2})$ | $34.934\ (94)^b$ | $10.115$ | $(27)^b$ |

$^a$ McAlexander et al. [267].
$^b$ One standard deviation.
$^c$ Systematic uncertainty.
$^d$ Jones et al. [194].
$^e$ Wang et al. [430].
$^f$ Freeland et al. [137].
$^g$ Rafac et al. [322].

this approach to study sodium and potassium, respectively. Both groups used the "pure" long-range $0_g^-$ state described by the Movre–Pichler Hamiltonian (Eq. 5.21) to obtain P - state lifetimes with uncertainties of around the 0.1% level. As in the case of lithium, at this level of precision retardation effects ([69], McLone and Power [269], Meath [270]) have to be taken into account. This means that in Eq. 5.21 the $C_3$ term has to be replaced by $f^{\Pi/\Sigma}C_3$ for the respective $\Pi$ and $\Sigma$ states, where $f^{\Pi/\Sigma}$ are the retardation correction factors,

$$f^\Pi = \cos\left(\frac{2\pi R}{\lambda}\right) + \left(\frac{2\pi R}{\lambda}\right)\sin\left(\frac{2\pi R}{\lambda}\right) - \left(\frac{2\pi R}{\lambda}\right)^2\cos\left(\frac{2\pi R}{\lambda}\right)$$
$$\simeq 1 - \frac{(2\pi R/\lambda)^2}{2} \tag{5.22}$$

and

$$f^\Sigma = \cos\left(\frac{2\pi R}{\lambda}\right) + \left(\frac{2\pi R}{\lambda}\right)\sin\left(\frac{2\pi R}{\lambda}\right) \simeq 1 + \frac{(2\pi R/\lambda)^2}{2}. \tag{5.23}$$

The significance of this factor can be gauged by comparing the magnitude of the retardation correction (121 MHz) to the experimental precision ($\pm 5$ MHz) for the case of $\text{Na}_2$ $0_g^-$ (Jones et al. [194]).

Table 5.1 summarizes the results on precision lifetime measurements. The currrent best lifetimes for Li, Na, K, and Rb are from photoassociation spectroscopy. Only the Cs lifetime is based on conventional methods (Rafac et al. [322]). The $C_3$ coefficients for the five attractive states from $^2\text{P}_{3/2} + {}^2\text{S}_{1/2}$ can be found by using the $d^2$ from Table 5.1 with the formulas in Table I of Julienne and Vigué [206].

## 5.7  Determination of the scattering length

As the kinetic energy of a binary collision approaches zero, the number of partial waves contributing to the elastic collision reduces to the s-wave, and the information inherent in the scattering event can be characterized by the scattering length, as defined by Eq. 2.20,

$$A_0 = -\lim_{k \to 0} \left( \frac{\tan \eta_0}{k} \right),$$ (5.24)

where $k = 2\pi/\lambda_{dB}$ is the wave vector, $\lambda_{dB}$ is the de Broglie wavelength of the reduced mass particle, and $\eta_0$ is the s-wave phase shift due to the scattering potential. The role of the scattering length in ultracold ground-state collisions is examined in detail in Section 2.2. The sign and magnitude of $A_0$ is crucially important to the attainment of Bose–Einstein condensation (BEC) in a dilute gas [83]. Binary collisions with positive scattering length lead to large, stable condensates while negative scattering lengths do not; and the rate of evaporative cooling of the gas, necessary to achieve critical densities and temperatures of the BEC transition, depends on the square of the scattering length (Ketterle and Van Druten [216]).

Photoassociation spectroscopy has played a pivotal role in determining accurate and precise values for the scattering length in collisions of Li, Na, and Rb, the three gases in which the quantum degenerate phase has been observed. The essential idea is to use PAS spectroscopy to probe the phase of the ground-state wavefunction with great precision. Two approaches have been most fruitful: (1) determination of the nodal positions of the ground-state wavefunction from the relative intensities of spectral features and (2) determination of the spectral position of the last ground bound state below the dissociation limit. The first approach takes advantage of the fact that PAS line intensities are sensitive to the amplitude of the ground-state wavefunction for each contributing partial wave near the Condon point of the transition, as indicated by Eq. 2.57. In particular, the slowly varying "Condon modulations" (Thorsheim et al. [385], Julienne [199]) have minima which reveal the position of the ground-state wavefunction nodes. A quantum scattering calculation tunes the model ground-state potentials until the relative spectral intensities in the calculation match those determined by the PAS experiment. The calculated phase shift then determines the scattering length $A_0$ for the model. Although all three alkalis yielded excellent results to this line of attack, each one presented unique features of PAS analysis. The second approach uses a two-photon PAS spectroscopy to locate the exact position of the last bound states. After the initial free–bound photoassociation transition a second probe laser stimulates the excited population downward to the last bound vibrational level of the ground-state. The detunings of the two probe lasers determine the spectral position of the last bound state, which in turn fixes the threshold phase of the ground-state potential within highly constrained limits. A quantum scattering calculation of the s-wave elastic phase shift then yields the scattering length. Both of these strategies have been used to good advantage.

### 5.7.1 Lithium

The Rice group led by R. G. Hulet determined the scattering length of collisions between two $^7$Li atoms in the $F = 2$, $M_F = 2$ state using the two-photon technique to determine the last bound vibrational level of the $^7$Li$_2$(a $^3\Sigma_u^+$ ) triplet ground state [3]. They set up

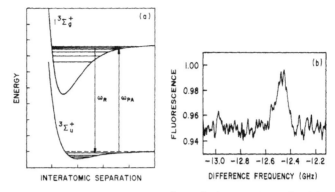

**Fig. 5.24.** Determination of the last bound state in $^7Li_2(a\,^3\Sigma_u^+)$ : (a) schematic of the stimulated excitation and de-excitation scheme; (b) peak in the MOT fluorescence as the second probe sweeps over the last vibrational level of the ground state [3].

a conventional lithium MOT and tuned a photoassociating laser to the $v = 64$ level of the $1\,^3\Sigma_g^+$ bound excited state, using trap-loss fluorescence to find the free–bound transitions. Then a *second* probe connected this bound excited state back to high-lying bound states of the $a\,^3\Sigma_u^+$ ground state. Resonances with the second probe were detected by an *increase* in the fluorescence base line, representing an anti-trap-loss fluorescence peak. Figure 5.24(a) shows the schematic of the stimulated excitation and de-excitation scheme, and Fig. 5.24(b) shows the peak in the MOT fluorescence as the second probe sweeps over the last bound vibrational level of the ground state. The difference frequency between the two probes, measured with a spectrum analyzer, permitted precise determination of the $a\,^3\Sigma_u^+$ last bound state energy with respect to the dissociation asymptote. Comparison of this energy with an accurate model potential identified the vibrational level with the $v = 10$ quantum number. Although the model potentials used to identify the last bound state (Moerdijk and Verhaar [284], Moerdijk *et al.* [283], Côté *et al.* [87]) were sufficient for that purpose, they both predicted a $v = 10$ binding energy of 11.4 GHz, whereas the Rice group directly measured $12.47 \pm 0.04$ GHz. The next step therefore was to modify the dissociation energy $D_e$ of the potential of Côté *et al.* so that the $v = 10$ level matched the measurement. The newly modified, fine-tuned model potential was then input to a quantum scattering calculation to determine the s-wave phase shift in the limit of zero temperature. Equation 5.24 then determined the scattering length. The final result, taking into account uncertainties in the long-range coefficients of the model potential, was found to be $A_T(2, 2) = -27.3 \pm 0.8$ a.u., where $A_T(2, 2)$ signifies the scattering length of the triplet potential starting from the $F = 2$, $M_F = 2$ doubly polarized atomic state.

In more recent work the Rice group used the Condon modulations in X $^1\Sigma_g^+ \rightarrow$ A $^1\Sigma_u^+$ photoassociation line intensities to determine the scattering lengths in $^6Li$ and $^7$ Li collisions [1]. Figure 5.25 shows a plot of the photoassociation line intensities for the three possible combinations of atomic hyperfine levels $(1 + 1, 1 + 2, 2 + 2)$ in the case of $^7Li$ collisions. The minima, corresponding to nodes in the ground-state wavefunction, occur at slightly different positions for each of the hyperfine asymptotes. Varying the model

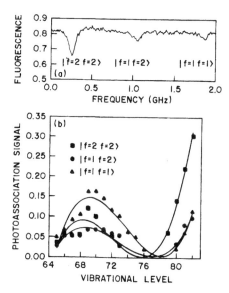

**Fig. 5.25.** Determination of the scattering lengths in $^7$Li collisions: (a) plot of the photoassociation line intensities for the three possible combinations of atomic hyperfine levels; (b) Condon fluctuations showing nodes occurring at slightly different positions for each of the hyperfine asymptotes, from [1].

potential of Côté *et al.* [87], until the scattering lengths give the best fit to the node positions, resulted in $A_S(1, 1) = 39 \pm 3$ a.u., and $A_S(2, 1) = 33 \pm 5$ a.u., and $A_S(2, 2) = 31 \pm 3$ a.u. for $^7$Li collisions. In the case of $^6$Li the ground-state splittings were too close to resolve minima from the different hyperfine hyperfine levels so the final result was reported as $A_S = 47 \pm 3$ a.u. for $^6$Li collisions. Finally, the Rice group completed the cycle in the lithium spectroscopy by determining the s-wave triplet scattering length in $^6$Li [2]. Once again, as in the case of the boson isotope $^7$Li, they used the two-photon photoassociation technique to determine the precise energy of a bound state ($v = 9$) of the a $^3\Sigma_u^+$ triplet ground state of $^6$Li$_2$. The result is also once again a negative scattering length, but of very large amplitude ($A_T = -2160 \pm 250$ a.u.), which indicates a near-threshold resonance lying just above the dissociation limit of the a $^3\Sigma_u^+$ state.

### 5.7.2 Sodium

The NIST group also used the relative intensities of their photoassociative ionization spectra to determine the scattering length in Na collisions (Tiesinga *et al.* [394]), but they were able to take advantage of a particularly fortuitous circumstance. One of the nodes of the ground-state p-wave (the "last" node before the wave assumes an oscillation characteristic of the 500 µK temperature) appears directly below $v = 0$ vibrational level of the pure long-range $0_g^-$ state. This p-wave node suppresses the intensity of photoassociation transitions to odd *rotational* levels of this $v = 0$ state and produces large variations in the rotational progression intensities for higher vibrational levels of the $0_g^-$ state as well. Figure 5.26

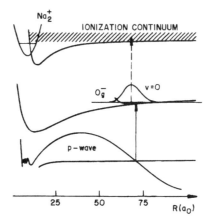

**Fig. 5.26.** Relative position of the ground-state p-wave, the $0_g^-$ state, and the ionization pathway, from Tiesinga, *et al.* [394].

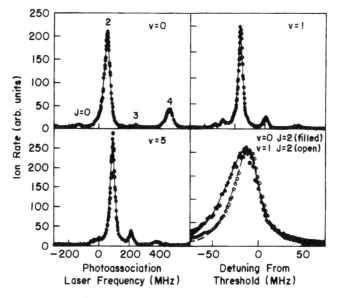

**Fig. 5.27.** Note how in Na PAI $(^3\Sigma_u \to 0_g^-)$ the $J = 1$ and $J = 3$ intensities are strongly suppressed in the $v = 0$ rotational progression because they arise from the p-wave contribution to the collision but then reappear in the $v = 1$ and $v = 5$ spectra, from Tiesinga *et al.* [394].

shows the relative positions of the ground-state p-wave, the $0_g^-$ state, and the ionization pathway. Figure 5.27 shows how the $J = 1$ and $J = 3$ intensities are strongly suppressed in the $v = 0$ rotational progression because they arise from the p-wave contribution to the collision but then reappear in the $v = 1$ and $v = 5$ spectra. By varying parameters in the singlet X $^1\Sigma_g^+$ and triplet a $^3\Sigma_u$ ground states, the NIST group was able to reproduce the relative intensities and line shapes of the PAI spectra. The fine-tuning of the potentials shifted the p-wave and s-wave nodes to positions of best-fit to the spectra, and from these nodal

**Table 5.2.** Scattering lengths ($a_0$): $A_s$, $A_t$ refer to singlet and triplet ground states, respectively. The remaining columns refer to $A_{f,-m_f}$

| | $A_s$ | $A_t = A_{f_{max}m_{max}}$ | $A_{1-1}$ | $A_{3,-3}$ | $A_{1,-1/2,2}$ |
|---|---|---|---|---|---|
| $^6\mathrm{Li}_2^a$ | $+45.5 \pm 2.5$ | $-2160 \pm 250$ | | | |
| $^7\mathrm{Li}_2^a$ | $+33 \pm 2$ | $-27.6 \pm 0.5$ | | | |
| $^6\mathrm{Li}^7\mathrm{Li}^a$ | $-20 \pm 10$ | $+40.9 \pm 0.2$ | | | |
| $^{23}\mathrm{Na}_2^b$ | | $+85 \pm 3$ | $+52 \pm 5$ | | |
| $^{23}\mathrm{Na}_2^{bb}$ | | | $+55.4 \pm 1.2$ | | |
| $^{39}\mathrm{K}_2^c$ | $+132 \leftrightarrow +144$ | | | | |
| $^{41}\mathrm{K}_2^c$ | $+80 \leftrightarrow +88$ | $+25 \leftrightarrow +60$ | | | |
| $^{39}\mathrm{K}_2^d$ | $+278 \pm 14$ | | | | |
| $^{40}\mathrm{K}_2^d$ | $+153 \pm 3$ | $+1.7 \pm 4.4$ | | | |
| $^{41}\mathrm{K}_2^d$ | $+121 \pm 2$ | $+286 \pm 36$ | | | |
| $^{39}\mathrm{K}_2^{dd}$ | | $-33 \pm 5$ | $-45 \pm 15$ | | |
| $^{41}\mathrm{K}_2^{dd}$ | | $+60 \pm 2$ | $+65 \pm 5$ | | |
| $^{85}\mathrm{Rb}_2^e$ | | $-500 \leftrightarrow -300$ | | | |
| $^{85}\mathrm{Rb}_2^{ee}$ | | $-520 \leftrightarrow -315$ | | | |
| $^{85}\mathrm{Rb}_2^f$ | $+4500 \leftrightarrow +\infty$ | $-440 \pm 140$ | | | |
| | $-\infty \leftrightarrow -1200$ | | | | |
| $^{87}\mathrm{Rb}_2^e$ | | $+85 \leftrightarrow +140$ | | | |
| $^{87}\mathrm{Rb}_2^{ee}$ | | $+101 \leftrightarrow +108$ | | | |
| $^{87}\mathrm{Rb}_2^g$ | | $+103 \pm 5$ | $+103 \pm 5$ | | $+103 \pm 5$ |
| $^{133}\mathrm{Cs}_2^h$ | $+280 \pm 10$ | $+2400 \pm 100$ | | | |
| $^{133}\mathrm{Cs}_2^i$ | | | | $|46 \pm 12|$ | |
| $^{133}\mathrm{Cs}_2^{ii}$ | | | | $+940$ | |
| $^{41}\mathrm{K}\ ^{87}\mathrm{Rb}^j$ | | $+163_{-12}^{+57}$ | | | |
| $^7\mathrm{LiH}^k$ | $+10 - 3 \leftrightarrow +10$ | $+65 \pm 5$ | | | |
| $^7\mathrm{LiH}^{kk}$ | | $+41.2$ | | | |
| $\mathrm{Mg}_2^l$ | $+26(9)$ | | | | |
| $\mathrm{He}_2^{*m}$ | | $+291$ | | | |

$^a$ Abraham *et al.* [2]. $^b$ Tiesinga *et al.* [394]. $^{bb}$ van Abeelen and Verhaar [408]. $^c$ Boesten *et al.* [49]. $^d$ Côté *et al.* [89]. $^{dd}$ Wang *et al.* [431]. $^e$ Boesten *et al.* [46]. $^{ee}$ Geltman and Bambini [148]. $^f$ Tsai *et al.* [401]. $^g$ Julienne *et al.* [203]. $^h$ Leo *et al.* [240]. $^i$ Monroe *et al.* [293]. $^{ii}$ Hopkins *et al.* [181]. $^j$ Ferrari *et al.* [132]. $^k$ Côté *et al.* [91]. $^{kk}$ Gadéa *et al.* [141]. $^l$ Tiesinga *et al.* [387]. $^m$ Gadéa *et al.* [142].

positions the scattering lengths for collisions between atoms in $F = 1$, $M_F = -1$ could be determined within narrow constraints. The result is $A_{1,-1} = 52 \pm 5$ a.u. Subsequent to the photoassociation work, van Abeelen and Verhaar [408] have used data from the magnetic field values of observed Feshbach resonances in Na collisions to constrain the $A_{1-1}$ scattering length to even tighter uncertainty limits. The results of their determination are given in Table 5.2.

### 5.7.3 Rubidium

The Texas group led by D. J. Heinzen, in collaboration with the Eindhoven theory group led by B. J. Verhaar, has carried out measurements of the scattering length in $^{85}$Rb collisions, using a FORT in which the atoms were doubly spin polarized (Gardner $et$ $al.$ [146]). The atoms approach only on the ground-state a $^3\Sigma_u^+$, and the photoassociation trap-loss fluorescence spectroscopy measures vibration–rotation progressions in the a$^3\Sigma_u^+ \rightarrow 0_g^-$ transition. Here the $0_g^-$ state is $not$ the pure long-range state but at short range becomes a component of the $^3\Sigma_g^+$ molecular state dissociating to $^2S_{1/2} + ^2P_{1/2}$. Figure 5.28 shows how the electron and nuclear spin polarization restrict the populated $0_g^-$ rotational states to even-numbered levels. Unlike the sodium case, with the $0_g^-$ state from $^2S_{1/2} + ^2P_{3/2}$, in which more than one partial wave contributed to various members of the rotational progression, the $0_g^-$ state from $^2S_{1/2} + ^2P_{1/2}$ has a one-to-one correspondence between the

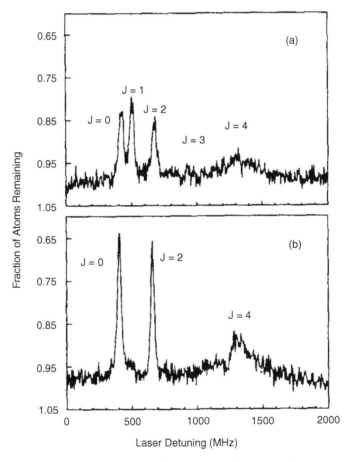

**Fig. 5.28.** Rubidium PAI spectra showing the effects of spin polarization: (a) Rb PAI spectrum unpolarized; (b) Rb PAI spectrum from doubly spin-polarized atoms. Note the restriction to even-numbered rotational levels, from Gardner $et$ $al.$ [146].

$J$ level and the scattering partial wave. The intensity of each $J$ peak effectively measures the amplitude of each partial wave. The energies of the $J = 0$ levels of five vibrational states were used to determine the product of the transition dipoles, $d(P_{1/2})d(P_{3/2})$, from a fit to the $C_3/R^3$ potential in the range of $R$ ($\sim$41–47 a.u.) where the photoassociation spectra are measured. A line-shape analysis similar to that of Napolitano *et al.* [302] was then carried out on the $J = 0$ and $J = 2$ features from which a triplet scattering length $-1000\, a_0 < a_T(^{85}\text{Rb}) < -60\, a_0$ was extracted. Using a $\sqrt{m}$ scaling procedure and a model triplet potential, the Texas group concluded that the triplet scattering length for $^{87}$Rb fell in the range, $+85\, a_0 < a_T(^{87}\text{Rb}) < +140\, a_0$. These measurements showed that Bose–Einstein condensation was not favorable for the triplet state of $^{85}$Rb but was possible for $^{87}$Rb. In a follow-up experiment on shape resonances in $^{85}$Rb collisions [49] the same group was able to narrow the range of triplet scattering length to $-500\, a_0 < a_T(^{85}\text{Rb}) < -300\, a_0$, confirming that $^{85}$Rb would not produce large, stable condensates.

### 5.7.4  Potassium

A joint experiment–theory effort by the Connecticut, JILA, and NIST groups [431] has determined the scattering length of the a $^3\Sigma_u^+$ triplet ground-state of potassium for various isotope combinations. The experiments used double-resonance photoassociative spectroscopy in a potassium MOT to measure transitions between the $0_g^-(v', J' = 2)$ pure long-range excited state and a series of molecular hyperfine levels belonging to the next-to-last vibrational level ($v'' = 25$) of the ground a $^3\Sigma_u^+$ state. The transitions were detected both by trap loss and by photoassociative ionization. Figure 5.29 shows the relevant potential curves and transitions for the two-photon photoassociation spectroscopy. The scattering length was calculated from the a $^3\Sigma_u^+$ potential resulting from a best fit to the measured molecular transitions. The results show that $^{39}$K, with a small negative scattering length, is not a good candidate for Bose–Einstein condensation; while $^{41}$K, with a reasonable isotopic abundance of 6.7% and positive scattering length is still a possibility.

### 5.7.5  Potassium-rubidium

Although the determination of the $a_3$ triple scattering length for the KRb molecule was not carried out by photoassociation spectroscopy, we describe the technique here and include the result in Table 5.2, which is a compilation of know scattering lengths for atomic species relevant to quantum degenerate gases.

The Italian group at Florence has carried out an experiment in which a mixture of cold $^{41}$K and $^{87}$Rb are trapped in a QUIC magnetic trap [126], and the $^{87}$Rb atoms are cooled by *forced evaporation* while the $^{41}$K atoms are *sympathetically* cooled by collisions with the Rb atoms [132]. The same technique was used to achieve BEC in $^{41}$K [281]. The Rb atoms are selectively heated by a 10% modulation of the magnetic trap at twice the frequency characteristic of the Rb atom radial oscillation. The measured rate of rethermalization through Rb–K elastic collisions determines the scattering length. Scattering lengths for other combinations of isotopes are determined by mass scaling.

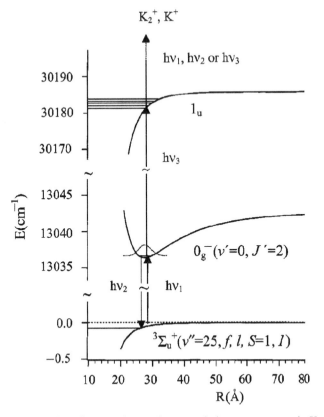

**Fig. 5.29.** Levels and transitions for two-photon photoassociation spectroscopy in $K_2$. ground-state atoms approaching in $f = 1$ or $f = 2$ hyperfine levels undergo photoassociation to the $0_g^-$ pure long-range state ($h\nu_1$) followed either by a subsequent spontaneous decay to $^3\Sigma_u^+(v'' = 25)$ ($h\nu_2$) or by two-photon ionization through a higher-lying molecular state ($h\nu_3$).

### 5.7.6  Lithium–hydrogen

The possibility of enhancing the Bose–Einstein condensation of hydrogen by using sympathetic cooling collisions with Li atoms was motivated Côté *et al.* to explore (via calculation) the scattering length of all the isotopic forms of this system [91]. This idea is appealing because the scattering length of hydrogen is known to be very small ($A_0 = 1.4\,a_0$, [216]), and the rate of evaporative cooling increases as the square of the scattering length. The basic idea of Côté *et al.* was to construct the best possible singlet ground-state potential for the LiH molecule by using an inverse perturbation analysis (IPA) of spectroscopic data, comparing the result to an *ab initio* calculation for the singlet potential, determining the fractional difference around the minimum, and then applying the same correction factor to the *ab initio* triplet potential. This procedure had the effect of deepening the triplet potential by 1.45 %. The procedure was then to numerically calculate the s-wave phase shift from these potentials, extrapolate the results to $T = 0$ and use Eq. 5.24 to find the scattering length. Côté *et al.* found that the triplet scattering lengths were large ($A_0 = 65$ $a_0$)and fairly stable with respect to uncertainties in the potential determination.

Gadéa *et al.* [141] questioned the procedure of inferring a correction to the *ab initio* triplet potential from the singlet potential behavior. They undertook to determine the scattering length for the LiH system using an entirely *ab initio* approach for the calculation of the triplet potential curve. They found significantly smaller values for the triplet scattering length ($A_0 \simeq +41$ $a_0$) but still large enough to predict a huge enhancement in the cooling rate of hydrogen by Li–H collisions.

### 5.7.7  Magnesium

Magnesium atoms have been cooled and trapped, and because of their closed-shell electronic structure, they exhibit no complicating hyperfine interactions. In order to investigate the scattering length for possible Bose–Einstein condensation, Tiesinga *et al.* [387] have carried out a calculation based on a constructed, state-of-the-art potential. At temperatures below 5 mK the ground-state collisions are entirely s-wave and the scattering length is calculated from a comparison of the asymptotic s-wave phase shift engendered by the potential to the freely propagating wave. The scattering length is positive and therefore an interesting candidate for Bose–Einstein condensation. Although this potential is not determined from photoassociation data, we include it here for the compilation in Table 5.2 and for its general interest.

### 5.7.8  Helium

Spin–polarized metastable helium atoms have been Bose condensed [310, 330] and the scattering length estimated to be $\simeq 350$ $a_0$ with uncertainties of $\pm 50\%$. Gadéa *et al.* undertook an *ab initio* calculation of the relevant $^5\Sigma_g^+$ potential in order to determine the s-wave phase shift. Instead of extrapolating numerically to zero as did [91], they employed a semiclassical expression [159] relating the phase shift to the scattering length. The result of the calculation is subject to large uncertainties because the depth of the potential well is very close to supporting a 15th bound state. Nevertheless the final result entered in Table 5.2 is in reasonable agreement with the estimates from the He BEC experiments.

Table 5.2 summarizes the s-wave scattering length determinations for various isotopes and states of the atom species measured or calculated to date. Most of the experimental determinations come from photoassociation data, but not all.

# 6

## Optical shielding and suppression

### 6.1 Introduction

Photoassociation uses optical fields to produce bound molecules from free atoms. Optical fields can also prevent atoms from closely approaching, thereby *shielding* them from short-range inelastic or reactive interactions and *suppressing* the rates of these processes. Recently several groups have demonstrated shielding and suppression by shining an optical field on a cold atom sample. Figure 6.1(a) shows how a simple semiclassical picture can be used to interpret the shielding effect as the rerouting of a ground-state entrance channel scattering flux to an excited repulsive curve at an internuclear distance localized around a Condon point. An optical field, blue detuned with respect to the asymptotic atomic transition, resonantly couples the ground and excited states. In the cold and ultracold regime particles approach on the ground state with very little kinetic energy. Excitation to the repulsive state effectively halts their approach in the immediate vicinity of the Condon point, and the scattering flux then exits either on the repulsive excited state or on the ground state. Figure 6.1(b) shows how this picture can be represented as a Landau–Zener (LZ) avoided crossing of field-dressed potentials. As the blue-detuned suppressor laser intensity increases, the avoided crossing gap around the Condon point widens, and the semiclassical particle moves through the optical coupling region adiabatically. The flux effectively enters and exits on the ground state, and the collision becomes elastic. The LZ probability that the particle remain on the ground state as it passes once through the interaction region is,

$$P_g = P_S = \exp\left(-\frac{2\pi\hbar\Omega^2}{\alpha v}\right), \tag{6.1}$$

where $\alpha$ is the slope of the difference potential evaluated at the Condon point,

$$\alpha = \left|\frac{d\Delta}{dR}\right|_{R_C} = \left|\frac{d[U_e(R) - U_g(R)]}{dR}\right|_{R_C}, \tag{6.2}$$

where $v$ is the local velocity of the semiclassical particle and $\Omega$ is the Rabi frequency of the optical coupling,

$$\frac{\Omega_{ge}}{2\pi} = 17.35 \text{ MHz} \times \sqrt{I \text{ (Wcm}^{-2})} \times d_{ge}(\text{a.u.}). \tag{6.3}$$

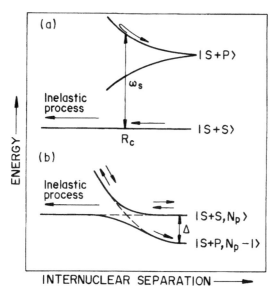

**Fig. 6.1.** Simple one-dimensional models of optical suppression: (a) conventional semiclassical picture of optical suppression. Atoms approach on the ground state and are excited in a localized region around $R_C$ by light of frequency $\omega_S$ to a repulsive excited state. (b) Dressed-state picture where asymptotic states are atom-field states and the optical coupling appears as a Landau–Zener avoided crossing.

In Eq. 6.3 $I$ is the shielding optical field intensity and $d_{ge}$ is the transition dipole moment in atomic units. Equation 6.1 expresses the probability of penetration through the crossing region with the scattering flux remaining on the ground state. This probability is often termed the "shielding measure," $P_S$. The probability of transferring to the excited state at the crossing is given by

$$
P_e = 1 - \exp\left(-\frac{2\pi\hbar\Omega^2}{\alpha v}\right),
\tag{6.4}
$$

and the probability of exiting the collision on the excited state asymptote after traversing the crossing region twice is

$$
P_e^\infty = P_e P_g = P_e(1 - P_e) = 1 - P_g^\infty.
\tag{6.5}
$$

According to Eqs. 6.1 and 6.5 the probability of the scattering flux exiting the collision on the ground state approaches unity as $I$ increases to the strong field regime. However, as we shall see in Section 6.2.4, other effects occur at high field to modify this simple behavior. In addition, such Landau–Zener models are necessarily phenomenological in nature, since there are a multiplicity of potential curves (see Fig. 4.9) and an effective value of the molecular transition dipole strength $d_{ge}$ must be chosen to fit the data.

## 6.2  Optical suppression of trap loss

Evidence for optical manipulation of trap-loss processes was first reported by Bali *et al.* [31] in $^{85}$Rb collisions in a MOT. In this experiment scattering flux entering on the ground state transfers to excited repulsive molecular states by means of a control or "catalysis" laser tuned 5–20 GHz to the blue of the cooling frequency. When the blue detuning of the control laser becomes greater than about 10 GHz, the scattering partners, receding from each other along the repulsive potential, gather sufficient kinetic energy to escape the trap. The control laser thus produces a "trap-loss" channel, the rate of which is characterized by a trap-loss decay constant, $\beta(I)$, which is a function of the control laser intensity, $I$. Figure 6.2 shows the increase of $\beta$ up to about 6 W cm$^{-2}$. Above this point the trap-loss rate constant begins to *decrease*, indicating a suppression of the loss channel with increasing control laser intensity. Bali *et al.* [31] interpret this suppression in terms of the dressed-state Landau–Zener crossing model depicted in Fig. 6.1. The suppression of $\beta$ is a consequence of the increasingly avoided crossing between the ground atom-field state, $|g; N\rangle$ and the excited atom-field state, $|e; N - 1\rangle$. As the control laser intensity increases, incoming flux on $|g; N\rangle$ begins to propagate adiabatically near the crossing and recedes along the same channel from which it entered. A follow-up study by this Wisconsin group (Hoffmann *et al.* [175]) used a blue-detuned laser to excite incoming collision flux population from the ground level to repulsive excited levels in an Rb MOT. The kinetic energy gained by the atoms dissociating along the excited repulsive potential probed the limits of capture velocity as a function of MOT parameters.

The Connecticut group of Gould and coworkers has also studied shielding and suppression in ground-state hyperfine-changing (HCC) collisions (Sanchez-Villicana *et al.* [337]). As discussed in Section 4.3, exoergic hyperfine-changing collisions can also lead to loss of atoms from a trap if it is made sufficiently shallow (by decreasing the trapping light intensity) such that the velocity gained by each atom from the HCC kinetic energy release exceeds the maximum capture velocity of the trap. In earlier studies, Wallace *et al.* [422]), observed

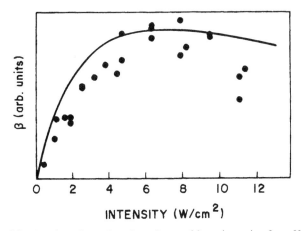

**Fig. 6.2.** Trap loss $\beta$ as a function of control laser intensity, from [31].

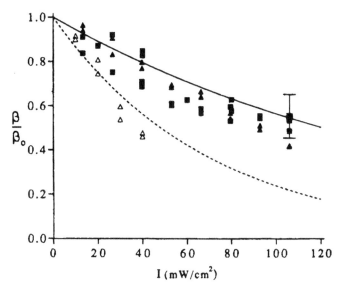

**Fig. 6.3.** Suppression ratio $\beta/\beta_0$ vs. suppression laser intensity $I$ for detunings of 500 MHz (solid points) and 250 MHz (open points). Different shaped symbols represent data from different runs. The solid and dashed lines show the Landau–Zener model calculations for the two detunings, from Sanchez-Villicana *et al.* [337].

a rapidly rising branch of $\beta$, the trap-loss rate constant, as a function of decreasing MOT laser intensity below about 2.5 mW cm$^{-2}$ in a $^{85}$Rb MOT and below about 4 mW cm$^{-2}$ in a $^{87}$Rb MOT. They interpreted this rapid rise as being due to HCC kinetic energy release. In order to suppress this source of trap loss Sanchez-Villicana *et al.* [337]) imposed a light field on their $^{87}$Rb MOT blue-detuned 500 MHz with respect to the $S_{1/2} + P_{3/2}$ asymptote. This suppressor light field couples the incoming ground-state collision flux to a repulsive excited-state potential at a Condon point $R_C \sim 32$ nm, effectively preventing the atoms from approaching any closer. The atoms recede, either along the excited-state potential or along the ground-state potential, before reaching the internuclear distance where HCC takes place ($\leq 10$ nm). The blue-detuned light field effectively shields the atomic collision from the HCC loss process by preventing the atoms from approaching close enough for it to take place. Figure 6.3 shows the effectiveness of this shielding action as a function of suppression intensity by plotting the ratio of trap loss rate constant $\beta/\beta_0$ with and without the suppressor laser being present. Sanchez-Villicana *et al.* [337] compare their measurements to the simple Landau–Zener model of the shielding effect (see Fig. 6.1 above) and find reasonably good agreement. The use of blue-detuned optical fields to suppress HCC has also been studied in cold sodium collisions. Muniz *et al.* [298] employed a novel sequential trapping technique to capture Na atoms in a MOT sufficiently shallow to observe the HCC loss effect. The Brazil group captured $10^7$–$10^8$ atoms in a relatively deep MOT operating near the D$_2$ line of Na. They then transferred these cooled atoms to a much shallower MOT operating near the D$_1$ line, the collisional loss from which is due principally to HCC. Then with a suppressor laser, blue-detuned 600 MHz with respect to the $S_{1/2} + P_{3/2}$

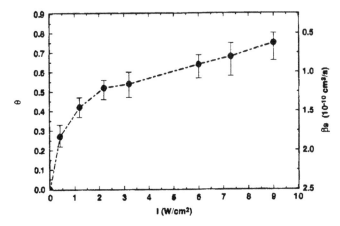

**Fig. 6.4.** Fractional suppression $\theta$ vs. suppressor laser intensity for a $D_1$ trapping laser intensity of 40 mWcm$^{-2}$, Muniz *et al.* [298].

asymptote, Muniz *et al.* [298] measured the suppression of HCC. Figure 6.4 shows a plot of the fractional decrease in HCC rate constant, $\theta = (\beta_0 - \beta)/\beta_0$ versus the suppression laser intensity, $I$. Both Sanchez-Villicana *et al.* [337] and Muniz *et al.* [298] compared their results to the simple Landau–Zener model, the diagram of which is shown in Fig. 6.1. From Fig. 6.3 it appears that the LZ model underestimates the suppression effect somewhat in $^{87}$Rb collisions, while Muniz *et al.* [298] find the model overestimates the effect in Na collisions. This model and its appropriateness to optical suppression and shielding is discussed in more detail in the next section. Muniz *et al.* [298] also compare their results to a more sophisticated three-dimensional, multichannel close-coupling calculation described in Section 6.2.4.

### 6.2.1   Optical shielding and suppression in photoassociative ionization

Although trap loss studies have played a major role in revealing the nature of cold and ultracold collisions, to date photoassociative ionization has been the only collisional process where the product is directly detected (although very recent results from Wang *et al.* [436] and Fioretti *et al.* [135] indicate direct observation of fine-structure-changing processes in K and Cs ultracold collisions, respectively). As discussed in Chapter 5 on photoassociation, the PAI process takes place in two steps. Reactant partners first approach on the ground state potential. Near the outer Condon point $R_C$, where an optical field $\hbar\omega_1$ couples the ground and excited states, population transfers to the long-range attractive state, which accelerates the two atoms together. At shorter internuclear distance, near an inner Condon point, a second optical field $\hbar\omega_2$ couples the population in the long-range attractive state either to a doubly excited state which subsequently autoionizes or couples directly to the ionization continuum. Optical shielding intervenes at the first step. Rather than transferring population to the long-range attractive potential at the first Condon point, an optical field $\hbar\omega_3$, tuned to the blue of the excited-state asymptote, transfers population to a repulsive curve from which

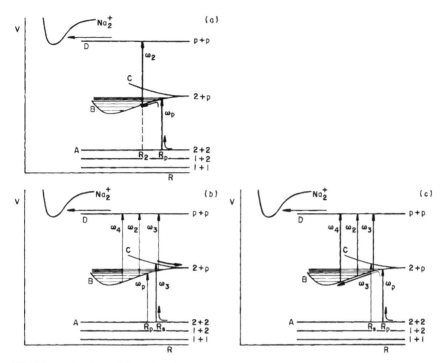

**Fig. 6.5.** Schematic of transitions showing PAI and optical suppression of PAI: (a) two-step PAI process; (b) suppressor frequency $\omega_3$ imposed on the collision, rerouting incoming flux to the repulsive excited curve; (c) with $\omega_p$ tuned to the right of $\omega_3$ PAI takes place with enhanced probability due to the addition of $\omega_3$ and $\omega_4$ to $\omega_2$, from Marcassa *et al.* [260].

it exits the collision on the excited asymptote with kinetic energy approximately equal to the blue detuning of $\hbar\omega_3$. This process is analogous to the weak-field, trap-loss mechanism described by Bali *et al.* [31]. As the intensity of the suppressor field increases, scattering flux begins to exit on the ground state with a negligible increase in kinetic energy, and the collision partners become *elastically* shielded from short-range interactions. Optical suppression of photoassociative ionization was first reported by Marcassa *et al.* [260] in a conventional MOT set-up. Four different optical frequencies were used to: (a) produce the MOT ($\omega_1$, $\omega_2$), (b) generate the suppressor frequency ($\omega_3$), and (c) probe the PAI inelastic channel ($\omega_p$). Electro-optic modulation of the repumper $\omega_2$ produced sideband frequencies $\omega_3$, $\omega_4$. Although $\omega_4$ was not necessary to observe the suppression effect, its presence added new features consistent with the interpretation of suppression and shielding described here. Figure 6.5(a) shows the coupling and schematic routing of the incoming flux. Figure 6.5(a) shows the familiar two-step PAI process with no suppression. After excitation to the attractive $2 + P$ level, the frequency $\omega_2$ transfers scattering flux in a second step to the $P + P$ level from which associative ionization takes place. As $\omega_p$ tunes to the red of the Condon point $R_2$, it can no longer, in combination with $\omega_2$, couple to the doubly excited level, and an abrupt cutoff of the PAI signal is predicted at that frequency. Figure 6.5(b) shows what happens when a suppressor frequency $\omega_3$ is imposed on the trap. With $\omega_p$ tuned to the *red* of

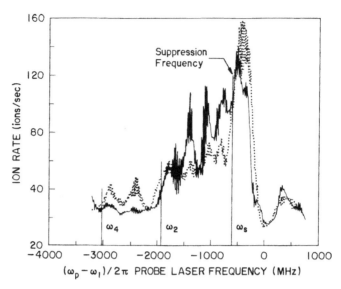

**Fig. 6.6.** Effect of optical suppression on the PAI spectrum observed in an Na MOT. The solid curve represents the PAI spectrum without the suppression field $\omega_s$ present. The dotted curve shows the PAI spectrum with $\omega_s$ present. Note the enhancement of the signal to the right of $\omega_s$ and suppression to the left. Note also the cutoff of the PAI signal to the left of $\omega_2$ without the suppression field and the extension of this cutoff to $\omega_4$ when $\omega_s$ is present, from Marcassa *et al.* [260].

$\omega_3$, scattering flux on the $F = 2$, $F' = 2$ entrance channel diverts to the *repulsive* $2 + P$ curve around the suppressor Condon point $R_s$ and exits via the excited or ground state, depending on the intensity of the suppressor field. Figure 6.5(c) depicts what happens when $\omega_p$ tunes to the *blue* of $\omega_3$. In this case the incoming flux transfers to the *attractive* $2 + P$ curve around the Condon point $R_p$. However, now $\omega_3$ and $\omega_4$ add to $\omega_2$ to enhance the probability of second-step excitation to the $P + P$ level. Therefore PAI should be enhanced with respect to the PAI rate represented in Fig. 6.5(a). Furthermore the cutoff to PAI should be extended to the left of $R_2$ because $\omega_4$ is the blue sideband of $\omega_2$, appearing 1.1 GHz higher frequency. Figure 6.6 shows a plot PAI count rate versus $\omega_p$ detuning with and without $\omega_3$ and $\omega_4$. All three predicted features, suppression, enhancement, and cutoff extension, are evident and appear where expected. The dependence of any shielding or suppression effect on the intensity of the optical field is obviously of great interest and importance. Marcassa *et al.* [260] described model two-state calculations at three levels of elaboration: (1) time-dependent Monte Carlo wavefunction simulations of wave-packet dynamics, (2) quantum close-coupling calculations without excited-state spontaneous emission, and (3) a very simple estimate based on a dressed-state Landau–Zener avoided crossing model. The result of the calculations were in close agreement among themselves and within about 15% of the experiment, given that an effective value of $d_{ge}$ in Eq. 6.3 had to be selected in order to give the model an intensity scale that could be compared with the experimental one. However, laser source limitations precluded intensity measurements above $\sim 1$ W cm$^{-2}$.

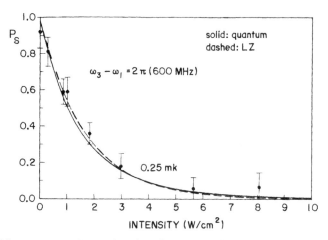

**Fig. 6.7.** Shielding parameter $P_s$ as a function of suppressor field intensity, Marcassa *et al.* [259].

In a follow-up article Marcassa *et al.* [259] were able to measure the suppression power dependence up to 8 W cm$^{-2}$. Figure 6.7 plots the fractional intensity of the PAI signal as a function of suppressor power density. The experimental results are shown together with calculations from quantum close coupling and the Landau–Zener semiclassical model. The agreement is good at low field values, but above about 5 W cm$^{-2}$ the data tend to lie above the theory curve, suggesting that the intensity dependence of suppression "saturates" at an efficiency of less than 100%. The calculated values do not lie outside the statistical error bars on the data points, however, so these data do not establish a "saturation" effect at high field beyond question. Zilio *et al.* [453] continued to investigate the high-field behavior of optical suppression of PAI, by measuring the rate dependence on the polarization (circular or linear) of the probe beam $\omega_p$. They found that as the suppressor field increased above about 1 W cm$^{-2}$, circular polarization became markedly more effective at suppressing PAI, but that neither linear nor circular polarization suppressed the PAI rate to the extent predicted by the simple weak-field LZ model. Figure 6.8(a) shows the shielding measure $P_S$ as a function of the suppressor field intensity, and Fig. 6.8(b) shows the results of a close-coupling, strong-field model calculation of the process (Napolitano *et al.* [301]). The model calculation shows the correct ordering of the polarization dependence and the magnitude of the shielding measure (which is considerably less than the LZ prediction), although the curvature of the experiment and theory plots have the opposite sign. This close-coupling calculation will be discussed in more detail in Section 6.2.4 on strong-field theories of optical shielding. In later work, Tsao *et al.* [405] investigated the alignment effect of linear polarization by measuring $P_S$ with the suppressor beam $E$-field aligned parallel and perpendicular to a highly collimated and velocity-selected atomic beam. They compared their results to the predictions of the close-coupling high-field model of Napolitano *et al.* [301] and found reasonably good agreement. Figure 6.9 summarizes the experimental result and the comparison with the model calculation.

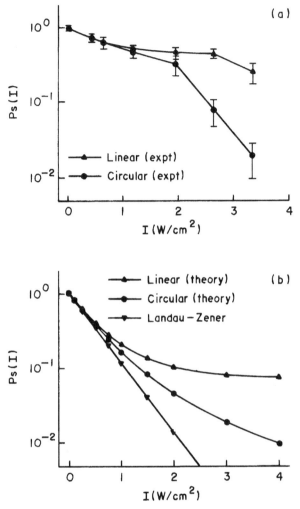

**Fig. 6.8.** Comparison of experiment and theory in optical suppression in Na: (a) experimental results showing the increased effectiveness of optical suppression by circularly polarized light at higher field intensities, from Zilio *et al.* [453]; (b) theory calculations in good qualitative agreement with experimental results. Also shown are the LZ results, which considerably overestimate suppression at high fields, from Napolitano *et al.* [301].

### 6.2.2   Optical shielding in xenon and krypton collisional ionization

Another example of optical shielding in ionizing collisions of metastable xenon has been reported by the NIST group (Walhout *et al.* [418]). The experiment takes place in a MOT with the cooling transition between the metastable "ground state," $6s[\frac{3}{2}]_2$, and the excited state, $6p[\frac{5}{2}]_3$. This intermediate coupling notation refers to the core total electronic angular momentum in square brackets, and the subscript refers to the total electronic angular momentum of the atom (core + outer electron). The MOT is time-chopped with a 150 μs

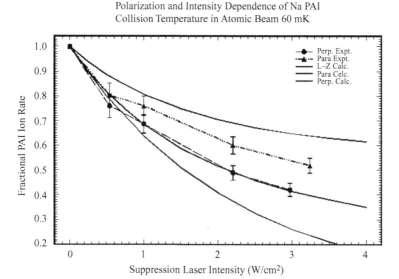

**Fig. 6.9.** Alignment dependence of optical suppression of Na PAI collisions in a beam. The results show the perpendicular polarization to be more effective than the parallel polarization and the LZ formula overestimates the effect at higher fields, from Tsao *et al.* [405].

period and an "on" duty cycle of $\frac{2}{3}$. Measurements were carried out in the probe cycle (MOT off) and the trap cycle (MOT on). Experiments in the probe cycle start with a control (or "catalysis") laser sweeping over a range of detuning from about 1.5 GHz to the red of the cooling transition to 500 MHz to the blue. The control laser induces a strong enhancement of the ionization rate over a red detuning range of about 500 MHz, peaking at 20 MHz, and produces suppression at blue detunings over several hundred megahertz. The maximum suppression factor (∼5), occurring at 200 MHz to the blue, appears to "saturate" at the highest control laser power of 0.5 W cm$^{-2}$. With the MOT beams illuminating the sample the experiments show an even more dramatic suppression effect. Figure 6.10 shows the suppression factor, the ratio of ionization rate constant with and without the control laser, plotted against control laser detuning. At about 250 MHz blue detuning the suppression factor reaches a maximum of greater than 30. The interpretation is similar to that of the sodium case (Marcassa *et al.* [260]. Without the control field present, the reactant scattering flux, approaching on the metastable ground state, is transferred to an attractive excited state at far internuclear separation by the trapping laser. The colliding partners start to accelerate toward each other and during their inward journey radiatively relax back to the metastable ground state. The consequent increase in kinetic energy allows the colliding partners to surmount centrifugal barriers that would otherwise have prevented higher partial waves from penetrating to internuclear distances where Penning and associated ionization take place. The net result is an increase in the ionization rate constant during the MOT "on" cycle. With the control laser present and tuned to the blue of the trapping transition, the previously accelerated incoming scattering flux is diverted to a repulsive excited state before reaching

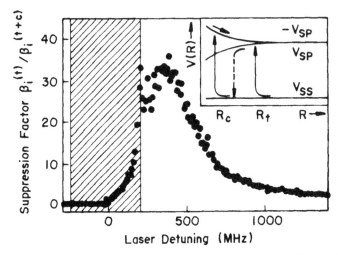

**Fig. 6.10.** Suppression factor as a function of control laser detuning in Xe collisions. These data show the suppression factor in the "trap" cycle with the MOT on, from Walhout *et al.* [418].

the ionization region. The colliding atoms are prevented from approaching further, and the result is the observed marked reduction in the ionization rate constant.

An experiment similar to the xenon work has also been carried out in a krypton MOT by Katori and Shimizu [212]. Analogous to the xenon case, the cooling transition cycles between $5s[\frac{3}{2}]_2 \leftrightarrow 5p[\frac{5}{2}]_3$ and a control laser sweeps from red to blue over the cooling transition from about $-600$ to $+100$ MHz. Suppression of the ionization rate was again observed with the control laser tuned to the blue. The power in the control laser was only about 4 mW cm$^{-2}$, and the suppression factor peaked at a blue detuning of about 20 MHz. Although this detuning is very close to the trapping transition, apparently the low power in the suppression laser and time chopping of the MOT cycle and probe cycle permitted ionization rate constant measurements without disruption of the MOT itself. Katori and Shimizu [212] also determined the ratio of rate constants between associative and Penning ionization and found it to be $\simeq 1/10$. The power dependence of the suppression effect was not investigated over a very wide range of control laser power. The maximum power density was only 25 mW cm$^{-2}$, so no firm conclusions concerning strong-field effects can be drawn from this report.

### 6.2.3  Optical shielding in Rb collisions

In other work Sukenik *et al.* [369] have observed evidence of optical suppression in two-photon "energy pooling" collisions in $^{85}$Rb and $^{87}$Rb MOTs. This study is a two-color experiment in which one frequency $\omega_1$ tunes 90 MHz to the red of the $5S_{1/2}(F = 1) \rightarrow 5P_{1/2}(F' = 1)$ $^{87}$Rb resonance transition (or the $5S_{1/2}(F = 2) \rightarrow 5P_{1/2}(F' = 2)$ $^{85}$Rb resonance transition) and populates singly excited attractive states originating from the $5S_{1/2} + 5P_{1/2}$ asymptote. A second color, $\omega_2$, tunes 0–2 GHz to the blue of $5S_{1/2}(F = 1) \rightarrow 5P_{1/2}(F' = 2)$ $^{87}$Rb transition (or $5S_{1/2}(F = 2) \rightarrow 5P_{1/2}(F' = 3)$ $^{85}$Rb transition). Violet photons are observed from the "energy pooling" energy transfer collision,

$$Rb(5P_{1/2}) + Rb(5P_{1/2}) \rightarrow Rb(5S_{1/2}) + Rb^*(6P) \tag{6.6}$$

followed by subsequent emission from the excited Rb atom,

$$Rb^*(6P) \rightsquigarrow Rb(5S) + hc/(\lambda = 421 \text{ nm}). \tag{6.7}$$

The spectrum of the violet photon emission was recorded as a function of $\omega_2$ detuning. Sukenik *et al.* [369] report deep modulations in the fluorescence intensity. They attribute this detuning modulation to transitions in which $\omega_2$ couples singly excited hyperfine levels, populated by $\omega_1$ excitation, to doubly excited molecular hyperfine levels. Studies of $\omega_1$ and $\omega_2$ intensity dependences show saturation in both colors. The $\omega_1$ saturation is due to depletion of the ground-state reactant flux. The $\omega_2$ saturation is ascribed to optical shielding due to the blue color coupling of the ground state and repulsive singly excited states. Thus $\omega_2$ is thought to play a dual role: (1) opening the energy-pooling channel by populating doubly excited molecular states at long range and (2) suppressing two-step photon transitions by coupling incoming ground-state flux to repulsive curves originating from the 5S + 5P asymptotes.

### 6.2.4  Theories of optical shielding

Weak-field theories applied to either Na or Xe experiments do not produce the experimentally measured dependence of ionization rate on blue-detuned optical field power density. In particular, a serious attempt to capture all the physics of strong-field optical shielding must take into account the effects of light shifts, saturation, rapid Rabi cycling at high fields, light polarization, and spontaneous decay. Suominen *et al.* [374] investigated these factors (except light polarization) within the simplifying context of two-state models in which the angular dependence of the suppressor field Rabi coupling was replaced by a spherical average. The essential motivation was to investigate the conditions under which the naive Landau–Zener formulas (Eqs. 6.1 and 6.4) break down and what modifications can be made, if any, to retain a semiclassical picture of the shielding dynamic. Suominen *et al.* [374] identify three different characteristic times, the relative duration of which determines the validity of the various approximations. These times are: (1) the lifetime against spontaneous decay, $\tau_\gamma$; (2) the time $\tau_{tp} \simeq \mu v/\alpha$ for a wave packet to travel from the excitation Condon point $R_C$ to the turning point $R_{tp}$ on the repulsive potential, where $\mu v$ is the local momentum; and (3) the time it takes a wave packet to traverse half the LZ interaction region, $\tau_\Omega \simeq (\hbar \Omega/2)/\alpha v$, where $v$, $\Omega$, and $\alpha$ have been defined in Eqs. 6.1, 6.2 and 6.3, respectively. Note that $\tau_\Omega$ is the only one of these characteristic times that depends on the shielding field intensity and reflects the increasing "repulsion" of the dressed potentials with increasing $\Omega$ at the avoided crossing. If $\tau_\gamma$ is long compared to both $\tau_{tp}$ and $\tau_\Omega$, spontaneous decay can be ignored and the weak-field LZ formulas are appropriate. If $\tau_\gamma$ becomes less than $\tau_{tp}$ but still remains long compared to $\tau_\Omega$, then spontaneous emission can be included by calculating the excitation probability with the LZ formula (Eq. 6.1) followed by radiative decay. In this case the shielding probability becomes,

$$P_S = P_g + P_e[1 - \exp(-\gamma \tau_{tp})] = 1 - P_e \exp(-\gamma \tau_{tp}). \tag{6.8}$$

If $\tau_\gamma$ becomes shorter than $\tau_{tp}$ *and* $\tau_\Omega$, then excitation and spontaneous emission cannot be separated because Rabi cycling is taking place around $R_C$. As the shielding field intensity increases, eventually $\tau_\Omega$ become greater than both $\tau_{tp}$ and $\tau_\gamma$, and the semiclassical particle or wave packet never passes entirely through the interaction region before undergoing Rabi cycling and spontaneous decay. In this high-field regime one would not expect the LZ approach to be very useful. Suominen *et al.* [374] investigated all of these relative time scales by comparing the LZ models to quantum Monte Carlo wave-packet calculations, the ultimate arbiter for the usefulness of any semiclassical two-level model. They found that, as expected, the LZ (Eq. 6.1) and modified LZ (Eq. 6.8) expressions fail with increasing $\Omega$ but that another modification of the LZ expression,

$$P_S = 1 - P_e \exp[-\gamma(\tau_{tp} - \tau_\Omega)] \tag{6.9}$$

actually tracks the Monte Carlo calculations quite well. The overall conclusion is that LZ and its modifications remain surprisingly robust even under conditions where one would expect them to fail, at least within the confines of a two-state system. Real collisions of course involve many more than two states, and the next question is to what extent off-resonant states can influence the conclusions of the two-state studies.

Suominen *et al.* [372] investigated this issue by carrying out Monte Carlo wavefunction (MCWF) calculations on a model three-state system consisting of a flat ground state and two excited states varying as $\pm C_3/R^3$. First, they investigated the effect of the off-resonant state (the repulsive $+C_3/R^3$ state) on the probability of inelastic trap-loss processes excited by a red-detuned laser resonant with the attractive $(-C_3/R^3)$ state at some Condon point $R_C$. The results showed that as the red-detuned laser Rabi frequency increased, the two-state and three-state MCWF calculations yielded essentially the same increasing probability of excited-state population while the simple weak-field LZ model predicted a low-level saturation of this excited-state population. Aside from the inclusion of the off-resonant state, the essential difference between the MCWF calculations and the LZ approach is that at high fields the MCWF method takes into account population recycling due to successive excitations after spontaneous emission. The LZ approach makes the weak-field assumption that excitation and spontaneous emission are decoupled so that after excitation to the attractive state, followed by spontaneous emission back to the ground state, re-excitation never occurs. Off-resonant population of the repulsive $+C_3/R^3$ state appears to have little effect on the results. Next, Suominen *et al.* [372] switched the roles of the attractive and repulsive excited states by investigating the influence of the off-resonant state (the attractive $-C_3/R^3$ state) on optical shielding when a blue-detuned laser is resonant with the repulsive $(+C_3/R^3)$ state at some Condon point $R_C$. They found that at weak field the MCWF results agreed with LZ calculations but that at strong field the shielding measure $P_S$ did not approach zero but began to *increase* due to off-resonant population of the attractive excited state. This behavior is not observed in the experiments of Zilio *et al.* [453] and Walhout *et al.* [418], where $P_S$ saturates with increasing shielding field intensity at levels above

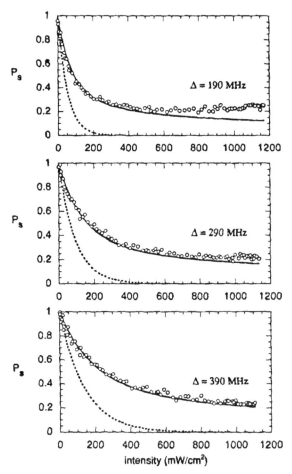

**Fig. 6.11.** Penetration measure $(1 - P_S)$ as a function of the laser intensity for three values of blue detuning of the control or shielding laser in Walhout *et al.* [418]. Circles represent experimental points, the solid line denotes the theory of Suominen *et al.* [372], and the dotted line denotes the standard Landau–Zener calculation.

those predicted by LZ but shows no evidence of a subsequent rise. The experimental results cannot be explained therefore by the presence of off-resonant states, but must be due to some other high-field effects.

At the same time that Suominen *et al.* [372] were carrying out the MCWF calculation to investigate the role of off-resonant states in a three-state model of shielding, they and their experimental coauthors (Suominen *et al.* [373] investigated to what extent a two-state Landau–Zener-type theory could explain the experimental results of Walhout *et al.* [418]. Both the alkali experiments (Maracassa *et al.* [260], Zilio *et al.* [453]) and the xenon experiments show that suppression and shielding are less than predicted by very simple two-state avoided-crossing models, but Fig. 6.11 shows that the extent of the shielding "saturation" as a function of shielding laser intensity is much more dramatic in the results of Walhout *et al.* [418]. The results of Fig. 6.11 were recorded with the MOT light turned off and only the blue-detuned control laser present. By inserting an averaged *distribution* of molecular Rabi frequencies in the Landau–Zener exponential argument, Suominen *et al.* [373] were able to get good agreement with these xenon experimental results. At first glance it might seem reasonable to use such a distribution since the MOT experiment does not restrict the angle $\theta$ between the collision axis and the control laser polarization axis, and the Rabi frequency could be written as $\Omega = \Omega_0 \sin \theta$. The problem is that there is no real justification for averaging $\Omega$ over $\theta$ since all three $P$, $Q$, $R$ branches are available for optical coupling. Furthermore, the experimental results show that *relative* shielding is much more pronounced when both the blue-detuned control laser and the red-detuned MOT lasers are turned on. With both red and blue colors present the extent of the shielding effect resembles the alkali results much more closely, and application of the LZ formula with a distribution of Rabi frequencies does *not* give good agreement. Although various *ad hoc* adaptations of the LZ formula can yield a satisfactory comparison with the xenon measurements, a clear, justifiable, and consistent physical picture does not emerge from these calculations.

Later Yurovsky and Ben-Reuven [452] have proposed an LZ approach in which the three-dimensional nature of the collision has been incorporated into the theory. By calculating the LZ probability at multiple crossings, where incoming $^1\Sigma_g$ ground-state s- and d-waves couple through $P$, $Q$, $R$ branches to a $^1\Pi_u$ repulsive excited state, several pathways are traced out through which the incoming flux can penetrate to the inner region, thereby rendering the shielding less efficient than predicted by a naive one-dimensional LZ model. Yurovsky and Ben-Reuven calculated the shielding measure $P_S$ as a function of shielding laser intensity for three detunings and compared to the measured results in xenon collisions (Suominen *et al.* [373]. Although the three-dimensional LZ theory better reflects the "saturation" behavior of $P_S$ than the one-dimensional LZ theory, the xenon measurements still show a stronger saturation than either the one- or three-dimensional versions of the two-state LZ model.

Motivated by the experimental results of Marcassa *et al.* [259], Zilio *et al.* [453] and Walhout *et al.* [418], Napolitano *et al.* [301] developed a full three-dimensional, close-coupled, quantum scattering approach to optical suppression of ultracold collision rates. This work differs from the MCWF approach of Suominen *et al.* [372, 374] in that the calculations are not restricted to two or three states or to only s-wave scattering. Although it does not take spontaneous radiation explicitly into account, Suominen *et al.* [374] have shown that at large enough blue detuning the decay of the upper state can be neglected, and Napolitano *et al.* [301] carried out their model calculations in a detuning regime where spontaneous emission can be safely ignored. The model collision examined is,

$$A(^1S) + A(^1S) + P(\varepsilon_q, \hbar\omega_L) \rightarrow A(^1S) + A(^1P), \tag{6.10}$$

where $A(^1S)$, $A(^1P)$ are atoms in the $^1S$ and $^1P$ ground and excited states, respectively, and $P(\varepsilon_q, \hbar\omega_L)$ represents a photon of energy $\hbar\omega_L$ and unit polarization vector $\varepsilon_q$ with $q = 0$ for linear polarization and $q = \pm 1$ for circular polarizations. In order to keep the number of channels to a manageable size, nuclear and electron spins were excluded; but the model does represent real scattering of the spinless alkaline earths (group IIA) and serves as a qualitative guide for understanding the behavior of the Na and Xe experiments. The problem is set up with the atoms far apart in an asymptotic field-dressed atomic basis within a space-fixed frame defined by $\varepsilon_q$. As the atoms approach, this space-fixed basis correlates to a molecular body-fixed basis through the unitary transformation of the symmetric top eigenfunctions; and these molecular basis states are then coupled by the radiation field. This coupling is normally localized around a Condon point $R_C$ defined by the blue detuning of the shielding laser field. The molecular state formed from a pair of $^1S$ ground-state atoms ($^1\Sigma_g^+$) and the states formed by the $^1S,\,^1P$ ground, excited pair ($^1\Sigma_u^+,\,^1\Pi_u$) constitute the four-state molecular basis ($^1\Pi_u$ is doubly degenerate). The goal is to calculate the probability that scattering flux incoming on the molecular ground state will: (1) penetrate to the inner region where reactive and inelastic processes can take place, (2) transfer and exit on the $^1\Pi_u$ excited state, or (3) elastically scatter on the ground state.

The results of the quantum close-coupling calculations show that the shielding parameter $P_S$ is markedly sensitive to both optical field intensity and to polarization, with circular polarization shielding more effectively than linear polarization. Napolitano et al. [301] interpret the polarization effect as the difference in the number of angular momentum branches ($P, Q, R$) through which the ground and excited states can couple. For example Fig. 6.12 shows that when two partial waves $l = 0, 2$ cross in the region of the Condon point, the angular momentum selection rules result in different coupling for linear and circularly polarized light. For linear polarization the entrance d-wave couples only through the $P$ and $R$ branches, while for circularly polarized light $P, Q, R$ branches couple. The result is that circular polarization results in more "avoidedness" at the crossing than does

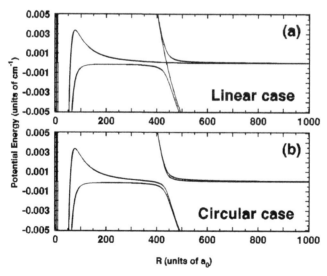

**Fig. 6.12.** Schematic of the close-coupling, strong-field model of optical suppression: (a) two partial waves, s and p, optically coupled from $^1\Sigma_g$ to $^1\Pi_u$, with linear polarization. Note that the d-wave couples only through $P, R$ branches. (b) The same coupling but with circular polarization. Coupling is through $P, Q, R$ branches, from Napolitano, et al. [301].

linear polarization. Figure 6.8, demonstrates a qualitative agreement between close-coupling calculations and the results of Zilio *et al.* [453] both in the amplitude of the shielding measure $P_S$ and in the ordering of the light polarization. A further result of the close-coupling theory is the prediction of anisotropy in shielding as a function of the linear polarization alignment with respect to the molecular collision axis. This behavior was confirmed in a beam experiment by Tsao *et al.* [405].

Most shielding and suppression studies, both experimental and theoretical, have focused on the dependence of $P_S$ as a function of optical field intensity at some fixed blue detuning. Band and Tuvi [34] have predicted, however, that by varying the relative detuning of the red and blue light fields in a sodium PAI experiment, as well as their intensities, one can produce interesting effects such as field-dressed "reaction barriers," the height and width of which can be controlled by the suppression field detuning and intensity. These barriers would be observable as resonances in the PAI signal as a function of blue detuning. The resonance positions should also shift and broaden with increasing suppression intensity. To date we are aware of no experimental efforts undertaken to investigate these interesting predictions of a model calculation, but shielding and suppression should provide new opportunities for optical control of inelastic processes.

# 7

# Ground-state collisions

Most of this review has focused on collisions of cold, trapped atoms in a light field. Understanding such collisions is clearly a significant issue for atoms trapped by optical methods, and historically this subject has received much attention by the laser cooling community. However, there is also great interest in ground-state collisions of cold neutral atoms in the absence of light. Most of the early interest in this area was in the context of the cryogenic hydrogen maser or the attempt to achieve Bose–Einstein condensation (BEC) of trapped doubly spin-polarized hydrogen. More recently the interest has turned to new areas such as pressure shifts in atomic clocks or the achievement of BEC in alkali systems. The actual realization of BEC in $^{87}$Rb [15], $^{23}$Na [103], $^{7}$Li [56, 57], $^{4}$He* [310, 330] and H [138] has given a tremendous impetus to the study of collisions in the ultracold regime. Collisions are important to all aspects of condensates and condensate dynamics. The process of evaporative cooling which leads to condensate formation relies on elastic collisions to thermalize the atoms. The highly successful mean field theory of condensates depends on the sign and magnitude of the s-wave scattering length to parameterize the atom interaction energy that determines the mean field wavefunction. The success of evaporative cooling, and having a reasonably long lifetime of the condensate, depend on having sufficiently small inelastic collision rates that remove trapped atoms through destructive processes. Therefore, ground-state elastic and inelastic collision rates, and their dependence on magnetic or electromagnetic fields, is a subject of considerable current interest.

This section will review work on ground-state collisions of trapped atoms in the regime below 1 mK, with particular emphasis on the ultracold regime below 1 μK. The work on ground-state collisions could easily be the subject of a major review in its own right, so our review will be limited in scope. We do not in any way claim to be exhaustive in our treatment. We will use a historical approach, as we have done for collisions in a light field, and try to cover some of the key concepts and measurements. The first section will review the early work in the field, including a brief survey of the work on hydrogen. A second section will discuss the role of collisions in BEC.

## 7.1 Early work

We noted in Section 2.2 that the quantum properties are quite well known for collisions where the de Broglie wavelength is long compared to the range of the potential. We confine our interest here to the special case of the collision of two neutral atoms at temperatures of less than 1 K. Interest in this subject was stimulated in the 1980s by two developments: the possibility of achieving Bose–Einstein condensation with magnetically trapped spin-polarized hydrogen [353, 354, 367], and the prospects of unparalleled frequency stability of the cryogenic hydrogen maser [93, 173, 184, 425]. The ground-state hydrogen atom has a $^2$S electronic state and a nuclear spin quantum number of $\frac{1}{2}$. Coupling of the electron and nuclear spins gives rise to the well-known $F = 0$ and $F = 1$ hyperfine levels of the ground-state. The transition between these two levels is the hydrogen maser transition, and the doubly spin polarized level, $F = 1$, $M = 1$, with both electron and nuclear spins having maximum projection along the same axis, is the one for which BEC is possible in a magnetic atom trap. Both the hydrogen maser and the phenomenon of BEC are strongly affected by atomic collisions of ground-state hydrogen atoms. Collisions cause pressure-dependent frequency shifts in the maser transition frequency that must be understood and controlled [94, 220, 221, 414], and they cause destructive relaxation of the spin-polarized H atoms that can prevent the achievement of BEC [10, 79, 358, 410].

These developments stimulated in the 1980s theoretical calculations for low-temperature collision properties of atomic H and its isotopes. Earlier work [11, 12, 96] had laid the groundwork for understanding inelastic spin exchange collisions by which two H atoms in the $F = 1$ state undergo a transition so that one or both of the atoms exit the collision in the $F = 0$ state. Berlinsky and Shizgal [42] extended these calculations of the spin exchange cross section and collisional frequency shift of the hyperfine transition to the low-temperature limit. These early calculations were based on extending the high-temperature theory, based on knowing the phase shifts of the ground-state molecular hydrogen $^1\Sigma_g$ and $^3\Sigma_u$ potentials alone without explicit inclusion of the hyperfine structure of the separated atoms. A proper quantum mechanical theory based on numerical solution of the coupled channel Schrödinger equation for the atoms with hyperfine structure, also known as the close coupling method, was introduced by the Eindhoven group, and applied to frequency shifts in hydrogen masers [221, 250, 414] and relaxation of doubly spin-polarized hydrogen in a magnetic trap [10, 234]. The close coupling method is a powerful numerical tool and is the method of choice for quantitative calculations on ground-state collisions. It is the best method currently available and has been applied to a variety of species, including mixed species. A recent discussion of the multichannel scattering theory for cold collisions has been given by Gao [145].

Collisions of species other than hydrogen have been investigated. Uang and Stwalley [406] looked at collisions of hydrogen and deuterium in cold magnetic traps to assess the role of deuterium impurities in a cold spin-polarized hydrogen gas. Koelman *et al.* [222, 223] calculated the lifetime of a spin-polarized deuterium gas, and Tiesinga *et al.* [386, 388] examined frequency shifts in the cryogenic deuterium maser. More recently Jamieson *et al.* [189] have calculated collisional frequency shifts for the 1S–2S two-photon transition in hydrogen. Tiesinga *et al.* [388] calculated that the relaxation rate coefficient for doubly spin-polarized Na would be about ten times larger than for spin-polarized hydrogen. Tiesinga *et al.* [392] also calculated that the frequency shift in a cesium atomic fountain clock (Kasevich *et al.* [75, 210] might be large enough to limit the anticipated accuracy of such a clock. The use of neutral atoms in ultra-precise atomic clocks is discussed by Gibble and Chu [153, 154], who measured large collisional frequency shifts in a Cs fountain. These have been verified by another experiment [151]. Verhaar *et al.* [413] argue that other cold collision properties can be deduced from these clock shift measurements, namely, that doubly spin-polarized $^{133}$Cs probably has a large negative scattering length; they criticize the opposite conclusion drawn from the same data by Gribakin and Flambaum [159] due to restrictive approximations used by the latter. Gibble and Verhaar [155] suggest that clock frequency shifts might be eliminated by using $^{137}$Cs. Kokkelmans *et al.* [228] use accurate collisional calculations for both isotopes of Rb to suggest that a Rb atomic clock will offer better performance than a Cs one. Gibble *et al.* [152] have directly measured the s-wave scattering cross section and angular distribution in Cs atom collisions in an atomic fountain experiment at a temperature of $T = 0.89\,\mu$K.

## 7.2 Bose–Einstein condensation

One of the major motivating factors in the study of collisions of cold ground-state neutral atoms has been the quest to achieve Bose–Einstein condensation. This is a phase transition which occurs in a gas of identical bosons when the phase space density becomes large enough, namely, when there is about one particle per cubic thermal de Broglie wavelength. The specific criterion [238] for condensation is

$$n\left(\frac{2\pi\hbar^2}{m\kappa T}\right)^{\frac{3}{2}} > 2.612, \tag{7.1}$$

where $n$ represents the atom density. In the condensate, there is a macroscopic occupation by a large number of atoms of the single ground-state of the many-body system, whereas in a normal thermal gas many momentum states are occupied with very small probability of occupying any given one. Achieving BEC means making the density large enough, or the temperature low enough, that Eq. 7.1 is satisfied. The early work with spin-polarized hydrogen aimed at reaching high enough density using conventional refrigeration techniques. Unfortunately, this proved to be impossible due to the losses caused by collisions or surface recombination when the density was increased. An alternative approach was developed using a magnetic trap without walls [174] and evaporative cooling [265] to reach much lower temperatures. The idea is to keep the density sufficiently low to prevent harmful collisions.

The success of laser cooling for alkali atoms gave an impetus to achieving BEC using alkali species. These are similar to hydrogen in that they have $^2$S ground electronic states with two hyperfine levels due to the nuclear spin (see, for example, Fig. 2.1). All alkali species have an isotope with odd nuclear spin which, together with the unpaired electron spin outside a closed shell electron configuration, results in a total integer spin, making the atom a composite boson. Unlike hydrogen, the alkali dimer $^3\Sigma_u$ state supports bound states; however, a metastable monomer condensate is possible because of the long time scale required to make dimer bound states *via* three-body recombination at low condensate density. Ordinary laser cooling methods produce density and temperature conditions many orders of magnitude away from satisfying the phase space density criterion in Eq. 7.1. The process of evaporative cooling was seen as a viable route to reaching BEC, and several groups set out to make it work. This approach was spectacularly successful, and within a few months of each other in 1995, three groups reached the regime of quantum degeneracy required for BEC. The first unambiguous demonstration of BEC in an evaporatively cooled atomic gas was reported by the NIST/JILA group[15] for doubly spin polarized $^{87}$Rb, followed by evidence of BEC for doubly spin-polarized $^7$Li by a group at Rice University [57], and then a demonstration of BEC at MIT [103] in the $F = 1, M = -1$ lower hyperfine component of $^{23}$Na. A much clearer demonstration of BEC for the $^7$Li system was given later [56]. In 1997, at least three additional groups used similar evaporative cooling and trap designs to achieve BEC in $^{87}$Rb and $^{23}$Na. In 1998 Bose–Einstein condensation was finally achieved in hydrogen. The detection signature was a large frequency shift in the two-photon transition from the 1S ground-state to the 2S excited state[138]. In 2000 the JILA group succeeded in obtaining a BEC in $^{85}$Rb [84], which was a real tour de force since the condensate was achieved by reversing the sign of the scattering length from negative to positive in the vicinity of a Feshbach resonance (see below). Then in 2001 two French groups [310, 330] independently produced a Bose condensate in the metastable $2\,^3S_1$ state of helium. These helium experiments had been inspired by an early suggestion of Shylapnikov *et al.* [351] that the metastable state of He might be a promising candidate. This long-lived species can be cooled and trapped [36, 129]. Fedichev *et al.* [129] calculated that the collisional ionization rate coefficient is so small for the $J = 1, M = 1$ level that a polarized gas of such a species might be stable long enough to make condensation possible. The polarized gas is stable because a collision of two $j = 1, m = 1$ atoms only occurs on the $^5\Sigma_g$ potential of the He$_2$ dimer, for which Penning ionization is forbidden. A gas of metastable atoms is only stable if complete polarization is maintained. An unpolarized sample would rapidly destroy itself due to very fast Penning ionization collisions [36]. The Penning ionization rate coefficient for other spin-polarized $j = 2$ metastable noble gases is not known. Experiments on cold, trapped $j = 2, m = 2$ Xe metastable atoms indicate that collisional ionization of the polarized gas is comparable to that for the unpolarized gas (Rolston [334]).

An Italian group [281] reported the achievement of BEC in $^{41}$K by sympathetic cooling in the presence of $^{87}$Rb. The sympathetic cooling technique is very promising for achieving quantum degeneracy in fermion systems where the Pauli exclusion principle prohibits efficient evaporative cooling. A Fermi gas, $^6$Li, has been immersed and thermalized in a BEC of $^7$Li [343] and a BEC of $^{23}$Na [161]; these mixtures provide the starting point for studying degenerate Bose–Fermi quantum gas mixtures. Although degenerate fermion gases should reveal interesting physics, the Fermi statistics prohibit s-wave (or any even-wave) scattering. The lowest odd partial wave that contributes to a collision cross section is the p-wave, and at BEC temperatures (on the order of tens or hundreds of nanokelvins) the p-wave centrifugal barrier will prevent Fermi gas atoms from approaching closely.

A large literature on the subject of BEC has already been generated. Some introductory articles on the subject are given by [60, 83, 398]. Later reviews on cold collisions in BEC [61] and progress in degenerate quantum gas physics [18] follow up the introductory articles.

## 7.3  Collisional aspects of BEC

Ground-state collisions play a crucial role both in the formation of a condensate and in determining its properties. A crucial step in the formation of a condensate is the achievement of critical density and temperature by "evaporative cooling" [172], the process by which hot atoms are removed from the confined ensemble while the remaining gas thermalizes to a lower temperature. Elastic collisions are necessary for evaporative cooling to work, and the stability and properties of the condensate itself depend on the sign and magnitude of the elastic scattering length. Two- and three-body inelastic collisions cause destructive processes that determine the condensate lifetime. Therefore, these collisions have been of as much interest for alkali species as for hydrogen and have been the object of numerous experimental and theoretical studies.

First, the process of evaporation depends on elastic momentum transfer collisions to thermalize the gas of trapped atoms as the trapping potential is lowered. These elastic collisions represent "good" collisions, and they have cross sections orders of magnitude larger for alkali species than for hydrogen. During evaporation, the rate of inelastic collisions which destroy the trapped hyperfine level, the so-called "bad" collisions, must remain much less than the rate of elastic collisions. An excellent description of the role of these two types of collisions is given in the review on evaporative cooling by Ketterle and van Druten [216]. Long before alkali evaporative cooling was achieved, Tiesinga *et al.* [389] calculated that the ratio of the "good" to "bad" collisions appears to be very favorable ($\sim$1000) for both the $F = 4, M = 4$ and $F = 3, M = -3$ weak-field-seeking states (i.e. those states confined by a magnetic trap) of $^{133}$Cs. Precise predictions were not possible due to uncertainties in the interatomic potentials. Monroe *et al.* [293] used time-dependent relaxation of trapped atoms to measure the elastic cross section for $F = 3, M = -3$ ground-state collisions near 30 $\mu$K. They measured a large value of $1.5(4) \times 10^{-12}$ cm$^2$ and found it to be independent of temperature between 30 and 250 $\mu$K. The measurements implied a scattering length magnitude near 46(12) $a_0$, if the cross section is assumed to be due to s-wave collisions. Newbury *et al.* [305] similarly measured the elastic cross section for the $F = 1, M = -1$ state of $^{87}$Rb to be $5.4(1.3) \times 10^{-12}$ cm$^2$, implying a scattering length magnitude of 88(21) $a_0$, as indicated in Table 5.2. Measurements on thermalization in an $F = 1, M = -1$ $^{23}$Na trap [102] deduced a scattering length of 92(25) $a_0$ for this level. A later study [19] of thermalization of doubly spin-polarized $^{133}$Cs $F = 4, M = 4$ showed that the scattering length magnitude was greater than 260 $a_0$, and the elastic scattering cross section was near the upper bound given by the s-wave unitarity limit, $8\pi/k^2$, between 5 and 50 $\mu$K. Another experimental study [181], this time of the absolute ground-state of Cs$_2$ (collisions on the $F = 3, M = -3$ asymptote), measuring rethermalization rates from nonequilibrium spatial and velocity distributions, in the temperature range from 1 to 30 $\mu$K and magnetic fields from 0.05 to 2.0 mT, inferred a lower bound on the s-wave scattering length of 940 $a_0$. All of these experimental studies measured only the total cross section, and *therefore were not able to determine the sign of the scattering length*. Verhaar *et al.* [413] and Drag *et al.* [116], however, calculated the sign to be negative for $^{133}$Cs $F = 4, M = 4$; but in light of new evidence from Feshbach resonance spectroscopy [72], Leo *et al.* [240] now calculate the $X^1\Sigma_g^+$ and $a^3\Sigma_u$ scattering lengths to be $(+280 \pm 10)\, a_0$ and $(+2400 \pm 100)\, a_0$, respectively. Therefore the achievement of BEC in $^{133}$Cs cannot be ruled out on the grounds of a negative scattering length. The scattering length for spin-polarized metastable helium has been calculated [142] using a large basis set to describe an *ab initio* $^5\Sigma_u^+$ potential. The determination of this scattering length is particularly sensitive to this potential because the depth is just near the threshold for supporting a 15th vibrational level, and the resulting scattering length will consequently have large amplitude but with uncertain sign. Gadéa *et al.* determine a positive scattering length of 291 $a_0$.

The second way in which atomic interactions profoundly affect the condensate properties is through their effect on the energy. The effect of atom–atom interactions in the many-body Hamiltonian can be parameterized in the $T \to 0$ limit in terms of the two-body scattering length [182]. This use of the exact two-body $T$-matrix in an energy expression is actually a rigorous procedure, and can be fully justified as a valid approximation [363]. One simple theory which has been very successful in characterizing the basic properties of actual condensates is based on a mean-field, or Hartree-like, description of the condensate wavefunction, which is found from the equation [123]:

$$\left( -\frac{\hbar^2}{2m} \nabla^2 + V_{trap} + NU_0|\Psi|^2 \right) \Psi = \mu\Psi, \tag{7.2}$$

where $V_{trap}$ is the trapping potential which confines the condensate and

$$U_0 = \frac{4\pi\hbar^2}{m} A_0 \tag{7.3}$$

represents the atom–atom interaction energy, proportional to the s-wave scattering length for the trapped atomic state, and $\mu$ is the chemical potential, i.e. the energy needed to add one more particle to a condensate having $N$ atoms. The condensate wavefunction in this equation, called the Gross–Pitaevski equation or the nonlinear Schrödinger equation (NLSE), can be interpreted as the single ground-state orbital occupied by each boson in the product many-body wavefunction

$$\Phi(1 \ldots N) = \prod_{i=1}^{N} \Psi(i).$$

The wavefunction $\Psi$ could also be interpreted as the order parameter for the phase transition that produces the condensate.

The effect of atom interactions manifests itself in the NLSE in the mean-field term proportional to the local condensate density, $N|\Psi|^2$, and the coupling parameter $U_0$ proportional to the s-wave scattering length $A_0$. In an ideal gas, with no atom interactions, $A_0 = 0$, and this term vanishes. Equation 7.2 shows that the condensate wavefunction in such an ideal-gas case just becomes that for the zero-point motion in the trap, that is, the ground state of the trapping potential. For typical alkali traps, which are harmonic to a good approximation, the zero-point motion typically has a frequency on the order of 50–1000 Hz, and a range on the order of 100 to a few μm. For comparison, $k_B T/h = 20$ kHz at $T = 1$ μK. In actuality, as atoms are added to the condensate the atom interaction term becomes the dominant term which affects the condensate wavefunction; and the shape of the condensate depends strongly on the size of the $U_0$ term, which is proportional to the product $N A_0$. The sign of the scattering length is crucial here. If $A_0$ is positive, the interaction energy increases as more atoms are added ($N$ increases) to the condensate. The condensate is stable and becomes larger in size as more atoms are added. In fact, a very simple approximation, called the Thomas–Fermi approximation, gives the condensate density $n$ by neglecting the kinetic energy term in Eq. 7.2 in relation to the other terms:

$$n = N|\Psi|^2 = \frac{\mu - V_{trap}}{U_0}. \tag{7.4}$$

This equation is remarkably accurate except near the edge of the trap where $\mu - V_{trap}$ approaches 0 or becomes negative, and describes condensates of $^{87}$Rb $F = 2$, $M = 2$ and $F = 1$, $M = -1$ and $^{23}$Na $F = 1$, $M = -1$, all of which have positive scattering length. Condensates of such species can be made with more than $10^6$ atoms. In contrast, if the scattering length is negative, as for $^7$Li, increasing the number of particles in the trap makes the interaction energy term in Eq. 7.2 become more negative. The condensate contracts as more particles are added, and in fact, only about 1000 atoms can be added to a $^7$Li condensate before it becomes unstable and can hold no more atoms, [41, 56, 111]. A condensate with negative scattering length is not possible in a uniform homogeneous gas, but in an atom trap the presence of zero-point motion does permit the existence of a very small condensate, as is the case for $^7$Li. There is an essential interplay between the quantized energy level spacings of the trap vibrational levels and the interparticle interaction [208]. As long as the trap spacings are greater than the collisional interaction energy, the condensate can exist. Since the total interaction energy is directly proportional to the number of atoms in the condensate, there is a critical number of atoms that can be added to the condensate after which the BEC becomes unstable. Evidently, small traps should be able to hold more identical boson particles with negative scattering length interaction than larger traps. The number of atoms that can be held in the trap $N_0$ can be expressed in terms of $N_{crit}$, the critical or maximum number of atoms, as

$$N_0 < N_{crit} \simeq \frac{\sqrt{\hbar/m\omega}}{|a|},$$

where the amplitude of the trap oscillations is given by $l_0 = \sqrt{\hbar/m\omega}$ and $|a|$ is the absolute value of the negative scattering length.

It is perhaps not obvious why a collisional property such as the scattering length determines the energetics of the interacting particles. A simple heuristic argument to indicate why this is the case can be given in relation to Fig. 2.3 in Section 2.2. The long-wavelength scattering wave has its phase shifted near $R = 0$ by the interaction potential. From the perspective of the asymptotic wave the effective origin of the oscillation near $R = 0$ is shifted by the presence of the potential, to $R = A_0 > 0$ for the case of positive $A_0$ and to $R = A_0 < 0$ for the case of negative $A_0$, as Fig. 2.3(b) shows. The kinetic energy associated with the long-wavelength asymptotic wave is affected by this shift in effective origin of oscillation. If one thinks of the two-particle system in terms of a single reduced-mass particle-in-a-box,

the left-hand wall of the box at $R = 0$ is moved to larger or smaller $R$, depending on the
sign of the scattering length. What is important for the energetics is whether the change
in energy is positive or negative, relative to noninteracting atoms (the box with a wall at
$R = 0$). Given that the energy of the ground-state of a reduced-mass particle of mass $m/2$
in a box of length $L$ is

$$E_{box} = \frac{\hbar^2}{m} \left(\frac{\pi}{L}\right)^2,$$

it is easy to work out that changing the length of the box from $L$ to $L + A_0$ changes the
energy by an amount proportional to $A_0/m$, thus lowering the energy for the case of negative
$A_0$ and raising it for the case of positive $A_0$. Another, complementary, way of looking at it is
to notice that the local curvature of the wavefunction is greater for "smaller" boxes (positive
scattering length), which can be interpreted as an increase in the local kinetic energy as
the two particles encounter each other at the internuclear distance where the asymptotic
form of the wavefunction must join the strongly interacting form. Larger boxes (negative
scattering length) result in less wavefunction curvature and consequently less local kinetic
energy. However, assigning local values for dynamical variables such as the kinetic energy
or momentum, based on local properties of the scattering wavefunction, only makes sense
within the WKB approximation. Use of the WKB picture in the "joining" region between
the inner and asymptotic parts of the wavefunction is a dubious practice. A rigorous analysis
gives the coupling term in Eq. 7.3.

### 7.3.1 Further comments on the scattering length

As Fig. 7.1 (Fig. 2.3 reproduced here for convenience) shows, the scattering length is related to the intercept of a straight line joining an inner rapidly oscillating region of the
wavefunction to the outer asymptotic region where the interaction potential is negligible
and the de Broglie wave is constant. The slope and the intercept of this straight line arise
from the asymptotic properties of the elastically scattered wave. As discussed by Joachain
[193], and set out in Section 2.2.1, these properties can be studied by first starting from the
radial Schrödinger equation

$$\left\{-\frac{\hbar^2}{2m}\left[\frac{1}{R^2}\frac{d}{dR}\left(R^2\frac{d}{dR}\right) - \frac{l(l+1)}{R^2}\right] + V(R) - E\right\} \mathcal{R}_l(k, R) = 0. \tag{7.5}$$

As in Section 2.2.1 we write the radial equation in terms of a "reduced" potential

$$U(R) = \frac{2m}{\hbar^2} V(R)$$

and the new set of functions

$$u_l(k, R) = R \mathcal{R}_l(k, R).$$

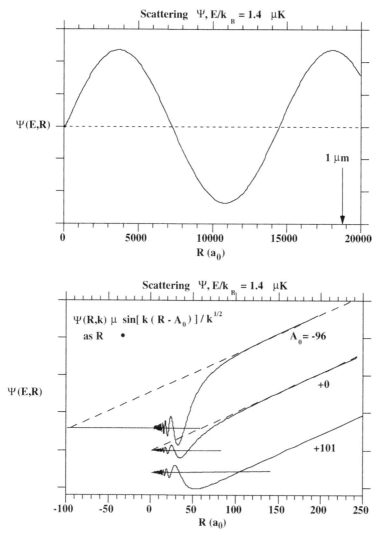

**Fig. 7.1.** The upper panel illustrates the long de Broglie wave at long range, on the scale of 1 μm. The lower panel shows a blowup of the short-range wave-function for the case of three different potentials, with three different scattering lengths, negative, zero, and positive.

The result is a simpler form for the radial equation

$$\left[\frac{d^2}{dR^2} + k^2 - \frac{l(l+1)}{R^2} - U(R)\right] u_l(k, R) = 0. \tag{7.6}$$

We consider only s-wave scattering and examine the form of the solution at low energy where $k \to 0$. In the long-range region where $U(R)$ is very small Eq. 7.6 becomes

$$\frac{d^2 u^0(R)}{dR^2} \simeq 0, \tag{7.7}$$

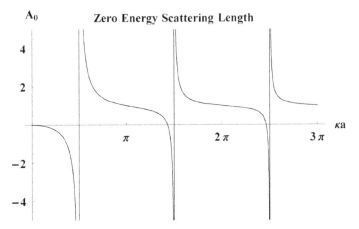

**Fig. 7.2.** Plot of the scattering length $A_0$ as a function of the phase shift $\eta(\theta)$ for a square-well potential. The scattering length is undefined at $\eta(\theta) = \pm (n + \pi/2)$, $n = 0, 1, 2, \ldots$. See the discussion in Section 2.2.2.

where $u^0(R)$ is the low-energy s-wave solution in this region. Clearly the form of the solution is

$$u^0(R) \simeq BR + C. \tag{7.8}$$

But we also know that in general the asymptotic form of the solutions to Eq. 7.6 is

$$u_l(kR) \rightarrow C_l(k) \sin\left[ kR - \frac{1}{2}l\pi + \eta_l(k) \right], \tag{7.9}$$

where $\eta(l)$ is the phase shift for the $l$th partial wave and $C_l$ is a normalization factor. For s-waves and low energy ($k \rightarrow 0$) the asymptotic solution takes the form

$$u_0(k, R) = C_0 \sin[kR + \eta_0(k)]$$

$$= C_0 k \cos \eta_0 \left( R + \frac{1}{k} \tan \eta_0 \right). \tag{7.10}$$

Equation 7.10 has the expected linear form where we identify the $R = 0$ intercept with the scattering length

$$A_0 \equiv - \lim_{k \to 0} \frac{\tan \eta_0}{k} \tag{7.11}$$

and we see that the slope is controlled by $\cos \eta_0$. Figure 7.2 shows the behavior of the scattering length as a function of the s-wave phase shift for a square-well potential. For zero phase shift, as Fig. 7.1 shows, the slope is such that the intercept goes through the scattering center, and the nonlinear term in the Gross–Pitaevski equation, Eq. 7.2, vanishes. For positive (negative) phase shifts, the slope diminishes and the scattering length is negative (positive). As the phase shift approaches $\pm(n + \pi/2)$, the slope approaches zero and the scattering length approaches to $\mp\infty$. The phase shifts at $\pm\pi/2$ correspond to *zero-energy resonances* where the incoming scattering state is just at the threshold of becoming

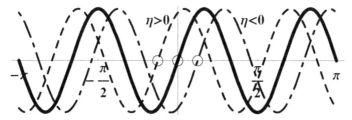

**Fig. 7.3.** plots of asymptotic scattered waves with different phase shifts. The solid curve shows zero phase shift, the dashed curve shows a positive phase shift, and the dot-dash curve shows negative phase shift. Note that a positive (negative) phase shift displaces the wavefunction closer to (further from) the scattering center.

bound. The appearance of zero-energy resonances in a square-well potential is discussed in Section 2.2.2. Exactly at the $\pi/2$ points the straight-line scattering wavefunction Eq. 7.10 exhibits zero slope and the intercept is at positive and negative infinity. The scattering length at these points is undefined.

Figure 7.3 shows the relative positions of the scattered waves with phase shifts positive and negative with respect to the incoming wave. From these phase shifts *and the scattering properties of the potentials giving rise to them* we can again offer a rationalization of why negative (positive) scattering lengths are bad (good) for BEC. A negative scattering length means that the scattered wave is shifted closer to the scattering center, and therefore the probability density of finding the two atoms closer together is enhanced. Equation 7.4 shows that the density increases which destabilizes the condensate. Since the long-range dispersive interaction is always attractive and the s-wave is the only one contributing to the collision (no orbital angular momentum), the atoms will tend to accelerate toward each other along their line of centers, and as the phase approaches $\pi/2$ the runaway density increase leads to collapse of the metastable condensate. In contrast a positive scattering length implies a scattered wave shifted outward from the center and a probability density favoring increased distance between the particles. The condensate wavefunction spreads out into a larger volume, raising its energy, decreasing its density, but making the condensate phase more stable against collapse. Repulsive and attractive potentials and the physical significance of the sign of the scattering length has also been discussed by Geltman and Bambini [148].

The gas thermalization studies which measure the elastic scattering cross section cannot determine the sign of the scattering length. The sign can be determined from photoassociation spectroscopy or Feshbach resonance spectroscopy, which are sensitive to the shape and phase of the ground-state collisional wavefunction. Much tighter constraints have also been placed on the magnitudes of scattering lengths by the photoassociation studies summarized in Chapter 5 above. Of course, observation of the condensate properties determines the sign as well. Even the magnitude of the scattering length can be found by measuring the expansion of the condensate when the trapping potential is removed, since the expansion rate depends on the strength of the interaction term in the initial condensate. This procedure has been carried out for both the $^{87}$Rb [178] and $^{23}$Na [70, 217, 273] condensates, confirming the magnitude of the scattering lengths determined more precisely by other means. Before the determinations by photoassociation spectroscopy, the sign and magnitude of the scattering lengths have also been estimated from the best available interatomic potentials based on spectroscopically derived potential wells and the long-range van der Waals potentials. This was done for $^7$Li [283, 284] and $^{23}$Na [284, 285], but the accuracy is not as good as for later determinations by photoassociation spectroscopy. Boeston et al. [49] have used spectroscopic data on K to calculate threshold scattering properties of $^{39}$K and $^{41}$K. Côté et al. [86] show that revised ground-state potentials for the $K_2$ molecule need to be used, and these give different scattering lengths from the calculation of [49]. The existence of photoassociation spectra for $K_2$ should permit a determination of more accurate threshold scattering properties of both isotopes in the future.

Another way in which collisions are significant for BEC is the role of inelastic collisions which change the state of the trapped species. This process can produce strong-field-seeking states (i.e. states ejected from a magnetic trap), as well as cause heating by releasing kinetic energy in the inelastic process. For example, if two doubly spin-polarized atoms collide and produce one or two atoms in the lower hyperfine state, then one or two units of the ground-state hyperfine splitting is given to the atoms, to be shared equally among them. Since this splitting is typically many hundreds of mK, the atoms easily escape the shallow traps designed to hold atoms at a few µK or less. We have already discussed how these "bad" collisions can affect the process of evaporative cooling. Fortunately, the ratio of "good" to "bad" collisions is favorable in many cases, so that evaporation actually works well and results in BEC. One case where evaporation appears to be unlikely due to "bad" collisions is the case of doubly spin-polarized $^{133}$Cs. Careful measurements [356] found that the inelastic rate coefficient for destruction of doubly spin-polarized $^{133}$Cs atoms is so large, about three orders of magnitude larger than predicted [389] that evaporative cooling of this level will be impossible. Leo et al. [239] explain this unusually large inelastic collision rate as being due to a second-order spin-enhancement of the effective spin-dipolar coupling. Evaporative cooling of the lower hyperfine level, $^{133}$Cs $F = 3, M = -3$ may still be feasible, if not wildly favorable. The collisional destruction rate for this level was measured in early work at JILA [293] and again by the French group led by J. Dalibard at l'Ecole Normale in Paris [160]. The JILA measurements indicate an upper limit on the dipolar relaxation rate of $>5 \times 10^{-14}$ cm$^2$ while the Paris results indicate a much larger cross section, $2 \times 10^{-13}$ cm$^2$. The Paris group notes that the dipolar relaxation rate for $^{133}$Cs $F = 3, M = -3$ is two orders of magnitude greater than for the lighter alkalis, making the prognostic for BEC in Cs less favorable than for the rubidium and sodium cases. Of course what really matters is the ratio of the elastic collision rate (good collisions) to the inelastic rate (bad collisions). The JILA measurement of the elastic rate at $1.5(4) \times 10^{-12}$ cm$^2$ [293] still indicates a favorable ratio of about 10. The sign of the scattering length for this level is still not certain.

Measuring an inelastic collision rate in a condensate, compared to the corresponding collision rate in a thermal gas, provides a way to probe the quantum mechanical coherence properties of the condensate [63]. Although the first-order coherence of a condensate can be measured by observing the interference pattern of two overlapping condensates [17], collisions probe higher-order coherence properties related to the different nature of the density fluctuations in a thermal gas and in a condensate. For example, for ordinary thermal fluctuations, the average of the square of the density is twice as large as the square of the average density, whereas for a condensate, these two quantities are equal. The second- and third-order coherence functions, $g^{(2)}(0)$ and $g^{(3)}(0)$, respectively, measure this effect, where the argument 0 implies that two or three particles are found at the same position; $g^{(2)}(0) = 2!$ and $g^{(3)}(0) = 3!$ for thermal gases, and both are 1 for a condensate. Thus atoms are bunched in a thermal source, but not in a condensate, analogous to photons in a thermal source and a laser. Kagan *et al.* [209] suggested that the three-body recombination rate might provide a way to measure this property, since the rate coefficient for three-body recombination would be $3! = 6$ times smaller in a condensate than in a thermal gas. The JILA group [63], in fact, has done just this, measuring $g^{(3)}(0) = 7.4(2)$ by comparing the measured three-body recombination rates for thermal and condensed $^{87}$Rb in the $F = 1$, $M = -1$ level.

Condensate coherence can also be probed with two-body inelastic collisions as well as three-body ones, as described by Ketterle and Miesner [217]. Stoof *et al.* [364] have used a collision theory viewpoint to show the difference of collision rates in a thermal gas and a condensate, showing that the corresponding rate for a two-body collision in a condensate is half that for the thermal gas. Reference [62] points out that a condensate will have a two-body photoassociation spectrum, which will also show the factor of 2 decrease in rate coefficient relative to a thermal gas. In addition, a condensate photoassociation spectrum would probe the two-body part of the many-body wavefunction in much more detail than an overall collision rate, since it probes the wavefunction over a range of interparticle separations instead of just yielding $g^{(2)}(0)$, as an overall collision rate does. Photoassociation should be readily observable in a condensate, since at the frequencies of photoassociation lines, the light absorption rate due to two-body photoassociation at typical condensate densities will greatly exceed the light scattering rate by free atoms. An interesting question concerning photoassociation in a condensate is whether a three-body spectrum could be observed, in which excited triatomic molecules are formed from three nearby ground-state atoms. If so, this could provide a means for a finer-grained probing of three-body effects in a condensate.

Inelastic collisions, including three-body recombination, are important for condensates and condensate formation because these can lead to heating or removal of atoms from the system. We discussed above how "bad" collisions can affect evaporative cooling. The lifetime of a condensate itself will be determined by collision processes. If the vacuum is not sufficiently low, hot background gas species will collide with trapped or condensed atoms, thereby transferring momentum to the atoms. Since the background gas atoms typically have energies on the order of 300 K, the cold atoms receive enough momentum to be ejected from the trap. There is also the possibility that some glancing collisions may transfer a very slight amount of momentum, producing hot but still trapped atoms. This could be a source of heating processes of unknown origin that have been observed in magnetic atom traps [273, 293]. Even if the vacuum is good enough, inelastic collisions among the trapped species themselves can limit the lifetime of the trapped gas or condensate. If the collision rate coefficient for the destructive process is $K_{in}^{(k)}$ for a $k$-body collision, the trap lifetime is $(K_{in}^{(k)} n^k)^{-1}$. For a nominal $10^{14}$ cm$^{-3}$ atom density in a condensate, a

1s lifetime results from a two-body rate coefficient of $10^{-14}$ cm$^3$ s$^{-1}$ or a three-body rate coefficient of $10^{-28}$ cm$^3$ s$^{-1}$. These rate coefficients will be very dependent on the species and the particular hyperfine level which is trapped. The only hyperfine levels for which trapping and condensation are possible are those for which the inelastic rate coefficients are sufficiently low.

We noted above how two-body inelastic collision rate coefficients must be small in relation to elastic collision rate coefficients in order for evaporative cooling to work. This is true for the $F = 2, M = 2$ $^{87}$Rb and $^7$Li species and $F = 1, M = -1$ $^{23}$Na species that have been condensed. The inelastic rate is small for the doubly spin-polarized species for the reasons discussed in Chapter 7 above. It is small for $F = 1, M = -1$ collisions for basically the same reason. An inelastic collision requires a weak spin-dipolar mechanism, since the sum of atomic M is not conserved, and additionally, the exit channel is a d-wave for an s-wave entrance channel. The small amplitude of the threshold d-wave leads to very small collisional destruction of $F = 1, M = -1$ for weak magnetic fields. The success of sympathetic cooling, and observation of a dual condensate of $^{87}$Rb $F = 1, M = -1$ and $F = 2, M = 2$ [299], raises the obvious theoretical question of why the inelastic collision rate coefficient for the destructive collision of these two species was found to be so small, $2.8 \times 10^{-14}$ cm$^3$ s$^{-1}$. This inelastic process, which produces two $F = 1$ hot atoms, goes by the spin-exchange mechanism, and normally would be expected to be several orders of magnitude larger. Three theory groups immediately answered the puzzle by showing that this observation meant that $^{87}$Rb had a very special property, namely the scattering lengths of both the $^1\Sigma_g$ and $^3\Sigma_u$ states are nearly the same, and in fact, all scattering lengths between any two hyperfine levels of $^{87}$Rb are nearly the same [59, 203, 226]. The existence of nearly identical scattering lengths is a sufficient condition for the inelastic rate coefficient to be as small as it is. It is not a necessary condition, since a threshold scattering resonance could also lead to a low inelastic rate coefficient. Such a resonance does not exist for $^{87}$Rb, however. Julienne *et al.* [203] pointed out that reconciling the existing data on $^{87}$Rb required that the scattering lengths for collisions between any two pairs of $F = 2, M = 2$ or $F = 1, M = -1$ differ by no more than 4 $a_0$ and have a value of $103 \pm 5$ $a_0$. Both Julienne *et al.* [203] and Kokkelmans *et al.* [226] calculated that the inelastic collisional destruction rate coefficient for collisions of $^{23}$Na $F = 2, M = 2$ and $F = 1, M = -1$ would be three to four orders of magnitude larger than that measured for $^{87}$Rb. Consequently a dual-species condensate would be impossible for $^{23}$Na. The issue of inelastic collision rate coefficients is crucial for the prospects of sympathetic cooling, which offers an attractive path for cooling species that cannot be cooled evaporatively, and needs to be investigated for mixed alkali species. For example, cooling is desirable for spin-polarized fermionic species such as $^6$Li, which may exhibit interesting Cooper pairing effects in the quantum degenerate regime [363]. Spin-polarized fermionic species cannot be cooled evaporatively, since only p-wave collisions are allowed, and these are strongly suppressed at low $T$ (see Section 2.2).

Two-body rate coefficients for inelastic processes tend to be small for species that can be condensed, since otherwise the bad collisions will limit the trap density. But as the density increases, three-body collisions will eventually provide a limit on the trap density and the lifetime. Three-body collisions produce a diatomic molecule and a free atom. These two products share the kinetic energy released due to the binding energy of the molecule. This is usually enough energy that the particles do not remain trapped; in any case, the molecule is unlikely to be trapped. Three-body collisions for spin-polarized hydrogen were studied by [104, 105]. The collision rate coefficient is unusually small for this system, since the ground-state triplet potential does not support any bound states in which to recombine, and making ground-state singlet molecules requires a very weak spin-dipolar transition. The rate coefficient for alkali systems is orders of magnitude larger than for hydrogen, since the triplet potentials support several bound states with small binding energy. Moerdijk et al. [282] calculate the three-body rate coefficients for doubly spin-polarized $^7$Li, $^{23}$Na, and $^{87}$Rb to be 2.6, 2.0, and 0.04 $\times 10^{-28}$ cm$^3$ s$^{-1}$, respectively. Moerdijk and Verhaar [286] differ from the suggestion of the JILA theory group [125] that the three-body rate coefficient may be strongly suppressed at low temperature. Fedichev et al. [129] give a simple formula based on the scattering length for the case when the last bound state in the potential is weakly enough bound. Burt et al. [63] note that this theory gives the magnitude of the measured three-body rate coefficient for $F = 1$, $M = -1$ $^{87}$Rb collisions.

### 7.3.2 Designer condensates using Feshbach resonances

One of the more interesting prospects for tailoring the collisional properties of ground-state species is to make use of an external field to modify the threshold collision dynamics and consequently to change either the sign or the magnitude of the scattering length or to modify inelastic collision rates. This prospect was raised by Tiesinga et al. [391] who proposed that threshold scattering properties for $^{133}$Cs could be changed by a magnetic field. This external control over the scattering length is possible because of the rapid variation in collision properties associated with a threshold scattering resonance. For example, a magnetic field can move a molecular bound state to be located at just the threshold energy for the collision energy of two levels of the lower hyperfine manifold. This sort of threshold phenomenon is called a *Feshbach resonance*. Figure 7.4 shows the basic mechanism of a Feshbach resonance. The scattering channel of an incoming s-wave overlaps at zero collision energy with a quasibound state associated with a molecular asymptote of higher energy than the entrance channel. When the magnetic moments of the quasibound state and the s-wave are different, their Zeeman shifts will differ; and a magnetic field can be used to tune the quasibound level onto scattering resonance. Moerdijk et al. [287] discussed the role of resonances for $^6$Li, $^7$Li, and $^{23}$Na. An initial experimental attempt [305] to locate predicted resonances in $^{87}$Rb was unsuccessful but there was really no doubt that such scattering resonances existed. The question was whether they could be found in experimentally accessible regimes of magnetic fields. Using much more refined calculations of threshold scattering derived from photoassociation spectroscopy, Vogels et al. [416] made specific predictions that resonances in scattering of the $F = 2$, $M = -2$ lower hyperfine level of $^{85}$Rb would occur in experimentally accessible ranges of magnetic field. A Feshbach resonance was indeed found by Heinzen's group at the University of Texas [92] at a roughly measured magnetic field of about 167 G, not too far from Vogels' prediction of 142 G. Soon after the JILA group, led by Wieman, reported observing and measuring the same

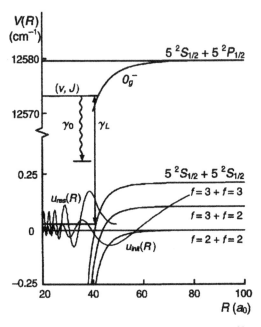

**Fig. 7.4.** Feshbach resonance in the case of ground-state scattering from [85] Rb atoms. The entrance channel wavefunction $u_{init}(R)$ couples to a quasibound sate with wavefunction $u_{res}(R)$. The energy level of the quasibound state is brought into a zero-energy resonance by tuning a magnetic field. Here the resonance is detected by enhancement of the photoassociation rate to a specific $v$, $J$ level of the $0_g^-$ excited state. $\gamma_L$.

resonance with greatly improved precision [332] from which they were able to determine new and improved values for the triplet and singlet scattering lengths. Soon after they were able to report even newer and more improved values [331]. The JILA group has also used this tunable scattering length to study instabilities in a [85]Rb BEC. By rapidly switching the scattering length from positive to negative values they have been able to trigger a system-atic and controlled collapse of the stable BEC [113, 333]. Figure 7.5 shows an image of an "exploded" BEC resulting from the sudden switching of the scattering length from an initial value $a_{init} = 0$ to negative values $a_{collapse}$ ranging from about $-2$ to $-50\,a_0$. The detailed kinetics of this collapse is not fully understood. There appears to be an induction time, after the switch to $a_{collapse}$ but prior to the actual collapse, which decreases strongly with the absolute magnitude of $a_{collapse}$. Theory has predicted [208] that immediately after the switch to negative scattering length the BEC should begin to contract, raising the energy and density until an instability point was reached. The instability is thought to be caused by three-body collisions which release sufficient energy to expel some the atoms from the BEC but not to destroy the BEC entirely [338]. However, the BEC density at the point of collapse, inferred from imaging the post-collapse expansion of the residual BEC, does not appear to contract markedly, and the details of the collapsing mechanism is at this point uncertain. The MIT BEC group, led by W. Ketterle [187] has also reported measurements of magnetically induced Feshbach resonance effects on condensate mean-field energy and

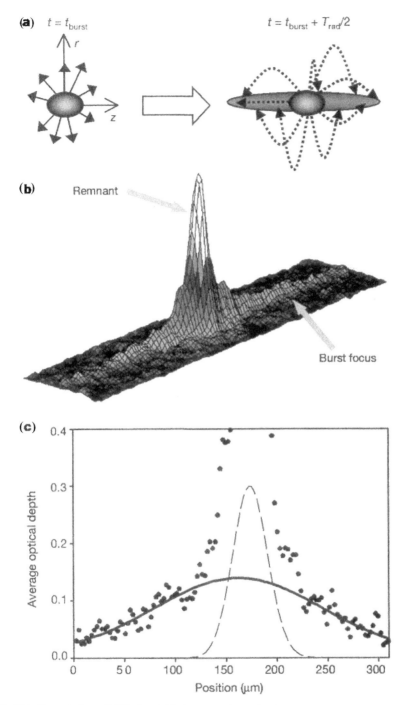

**(a)** $t = t_{burst}$     $t = t_{burst} + T_{rad}/2$

**(b)** Remnant

Burst focus

**(c)**

Average optical depth

Position (µm)

**Fig. 7.5.** This figure is from [113]. Panel (a) shows a schematic of a "burst focus" in which the exploding BEC is expelled radially outward by inelastic collisions with energy sufficient to destabilize a fraction of the BEC but without enough kinetic energy to escape the trap. The atoms all fall back onto the radial axis after a time $T_{radial}/2$, where $T_{radial}$ is the period of the radial harmonic oscillation of the trap. The fact that all the atoms reach the radial axis at the same time produces the "focus". Panel (b) shows an image of the radial burst together with BEC remanent sharply peaked at the center. Panel (c) is a fit of the radial extent of the burst which is a measure of the kinetic energy.

lifetime for Na $F = 1, M = +1$ confined in an optical trap. Boesten *et al.* [49] calculate that $^{41}$K may also offer good prospects for magnetic field tuning of the scattering length.

### 7.3.3 Nonmagnetic modulation of the scattering length

External fields other than magnetic ones can also change the scattering lengths. Fedichev and the Amsterdam theory group [128] proposed that an optical field tuned near resonance with a photoassociation transition can be used to vary the ground-state scattering length but not cause excessive inelastic scattering. Napolitano *et al.* [301] while investigating optical shielding (Chapter 6), calculated that large changes in elastic scattering rates can be produced by light detuned by about 100 natural line widths to the blue of atomic resonance. This was partly because the optically induced mixing with the ground-state of an excited repulsive excited state with the $1/R^3$ long-range form greatly changed the ground-state collision by inducing contributions from angular momenta other than s-waves. The problem with using light to change ground-state collision rates is that the light can also induce harmful inelastic processes as well, and the free atoms also scatter off-resonant light and experience heating due to the photon recoil. Such effects must be minimized in order for optical methods to be practical. Bohn and Julienne [50] considered the Fedichev proposal in more detail for the cases of $^7$Li and $^{87}$Rb, giving simple formulas for estimating the light-induced changes, and showed that there may be ranges of intensity and detuning where useful changes can be effected. Finally, the radio-frequency (rf) fields used in evaporative cooling can change collision rates. The Eindhoven theory group led by B. Verhaar [9] showed how the strong rf field in a microwave trap will modify collisions. Moerdijk *et al.* [288] showed that the rf fields used in evaporative cooling would make negligible changes in the collision rates of $F = 1, M = -1$ $^{23}$Na or $F = 2, M = 2$ $^{87}$Rb in a magnetic trap. Suominen *et al.* [375] agreed with this analysis for these species, but showed that typical rf fields for evaporative cooling can cause enhanced inelastic collisional relaxation of $F = 2, M = 2$ $^{23}$Na. This is because rf-induced nonadiabatic transitions due to motion in the trap lead to production of other $M$-levels, which decay with very large spin-exchange rate coefficients [203, 226].

### 7.3.4 Condensates in all-optical traps

Most of the experimental and theoretical effort in the study of quantum-degenerate gases has been directed toward magnetic confinement. The technology of magnetic traps has been well developed for quite some time [39] and the evaporative cooling of magnetically confined alkalis lead the way to the first condensates [216]. However, magnets, magnetic fields, and their rapid switching and sweeping can lead to undesirable transients in laboratory instrumentation. Furthermore, only "weak-field seeking" atomic states can be confined and great care must be taken to avoid nuclear spin flips leading to conversion of internal states to "strong-field seekers" and subsequent ejection from the trap. In some cases, such as $^{133}$Cs $F = 4, M = 4$ collisions, spin–spin relaxation is intrinsically very rapid; and the formation of BEC in weak-field seeking states appears to be highly unfavorable [160, 356].

An all-optical trap such as a dipole-gradient trap [320] would be desirable due to ease of switching and modulation, and because the trapping potential is not restricted to weak-field seekers. Furthermore, the trapping potential arises from the interaction of the optical-field-induced dipole with the field gradient. Therefore in principle any polarizable body (atoms, molecules, clusters) are susceptible to optical trapping without the requirement for a nonzero spin. As pointed out above, however, in Section 7.3.3 optical manipulation brings its own set of problems, most notably off-resonant absorption. The probability of absorption (or rate of scattering $\Gamma_{sc}$) falls off as the square of the detuning $\Delta$ from an isolated transition, and the optical trapping potential $U_{dip}(\mathbf{r})$ decreases only inversely,

$$U_{dip}(\mathbf{r}) = \frac{3\pi c^2}{2\omega_0^3} \left(\frac{\Gamma}{\Delta}\right) I(\mathbf{r})$$

$$\Gamma_{sc}(\mathbf{r}) = \frac{3\pi c^2}{2\hbar\omega_0^3} \left(\frac{\Gamma}{\Delta}\right)^2 I(\mathbf{r}).$$

Therefore, provided the atom light scattering is far from saturation, it should be possible to find an off-resonance regime where atoms can be optically trapped at an acceptably low absorption rate,

$$\hbar\Gamma_{sc} = \frac{\Gamma}{\Delta} U_{dip}.$$

The real challenge is to devise an absorption-free cooling scheme in an all-optical trap that will be sufficiently rapid to lead to BEC.

This challenge was first successfully met by Barrett *et al.* [37] where they achieved a BEC in $^{87}$Rb by confining and evaporating the atoms in a quasielectrostatic dipole force trap (QUEST) formed by two crossed $CO_2$ laser beams. The crossed beam geometry provides tight confinement in three dimensions and the $CO_2$ optical frequency ($2.828 \times 10^{13}$ Hz) is so far off resonance from the Rb resonance line ($3.844 \times 10^{14}$ Hz) that the absorption rate becomes negligible. They achieve sufficient evaporative cooling by starting with a large supply of precooled atoms from a magneto-optical trap (MOT), loading the MOT atoms efficiently into the QUEST and then lowering the QUEST potential by sweeping lower the $CO_2$ laser power over an interval of about 2 s. Prior to evaporative cooling the $^{87}$Rb atoms are optically pumped into the $F = 1$ hyperfine level. Since the QUEST is insensitive to nuclear spin, all three Zeeman sublevels of the $F = 1$ hyperfine level are Bose-condensed. The result is actually three co-existing BECs, one for each orientation of the nuclear spin. In the analysis of their evaporation kinetics, Barrett *et al.* posit that the principal loss process at early confinement times in the QUEST is due to three-body collisions. This process should heat the atoms, and the authors actually observe cooling during this early period so the source of this rapid, early loss is in some doubt.

Another series of experiments, developing an all-optical atom confinement near a surface, has been carried out by R. Grimm and his coworkers first at the Max Planck Institute for Nuclear Physics at Heidelberg and then at the Institute for Experimental Physics in Innsbruck. The first of these was the demonstration of a gravito-optical surface trap or GOST [309] in which they used an evanescent wave cooling technique to store an ensemble

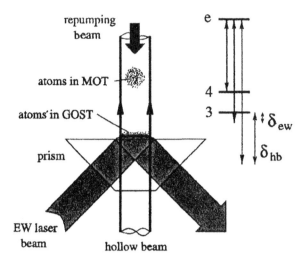

**Fig. 7.6.** The gravito-optical surface trap. Atoms are are loaded from a MOT. As the atoms fall they are transversely confined by the optical hollow beam, detuned to the blue with respect to $F = 3 \leftrightarrow e$ transition by an amount $\delta_{hb}$. The evanescent wave beam, blue detuned from the same transition by an amount $\delta_{ew}$, provides the repulsive potential against which the atoms bounce and collect at the bottom of the GOST.

of $10^5$ Cs atoms at about 2 µm above the surface of a dielectric prism with a temperature of 3 µK. This trap is not only magnet-free but is dissipative, i.e. can actively cool the confined atoms. A schematic diagram of the first version of the GOST is shown in Fig. 7.6. Owing to the large splitting between the two hyperfine ground-states of $^{133}$Cs, $F = 3$ and $F = 4$ ($\Delta E_{34} = 2\pi \times 9.193$ GHz), a kind of Sisyphus cooling cycle [108] makes this a dissipative trap, and atoms collect at the bottom with a temperature of about 3 µK. The cooling scheme works by first pumping the atoms into the $F = 3$ level, letting them drop and near the turning point at the bottom of the GOST, absorb a photon from the $F = 3$ level to the $F_e$ excited level. Spontaneous emission back to the $F = 4$ level will cool the atom by an energy quantum equal to $\Delta E_{34}$. As the atom bounces up, now in the $F = 4$ level, it is optically repumped to the $F = 3$ and the cycle is closed. Figure 7.7 shows a schematic of the cooling cycle. The atom densities obtained in this first version of the GOST were comparable to densities in a MOT ($\sim 10^{10}$ cm$^{-3}$; and, because the confined atoms are primarily in the lower hyperfine level, inelastic cold collisions did not significantly contribute to trap loss. In the next stage of development the Innsbruck group imposed a focused, red-detuned laser beam in the middle of the GOST, as depicted in Fig. 7.8, the purpose of which is to concentrate the atoms into a much smaller volume [163]. The Innsbruck group calls this technique a "dimple" trap because of the narrow attractive potential at the bottom of the GOST. It is reminiscent of the far-off resonance trap (FORT) introduced by Heinzen *et al.* [279] that consists of a far-red-detuned laser focused into a MOT. The novelty here is that the atoms confined in the "dimple" appear very close to ($\simeq 20$ µm) the dielectric surface. The atom density increases by a factor of about 300, approaching densities of $\sim 10^{14}$ cm$^{-3}$. At these densities two- and three-body collisional loss become significant. Further development in this series of

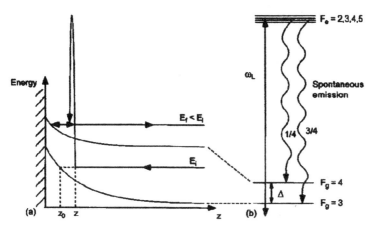

**Fig. 7.7.** Entrance and exit channel for atoms bouncing dissipatively in the GOST trap. Atoms enter
on the lower repulsive channel $F = 3$ and exit on the upper repulsive potential $F = 4$ after
absorption and spontaneous emission from the upper $F_e$ level. The atom kinetic energy is reduced
by $\Delta E_{34}$.

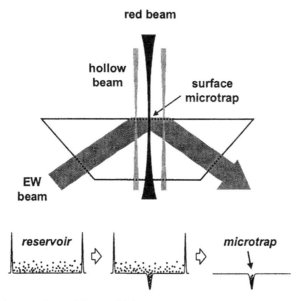

**Fig. 7.8.** Dimple microtrap formed from a GOST and a far-red-detuned focused laser aligned along
the GOST axis.

experiments is the dual evanescent wave trap (DEW trap) [162], which combines a blue-
detuned and a red-detuned evanescent wave to create a very tightly confined trap in the
vertical direction above the coupling prism. Figure 7.9 shows a sketch of the apparatus and
the resulting confining potentials. With this trap the Innsbruck group is at the threshold
of realizing a gas of reduced dimensionality. In contrast to the experiments with magnetic
confinement, constrained to the weak-field seeking $F = 4M = 4$ state, the all-optical DEW

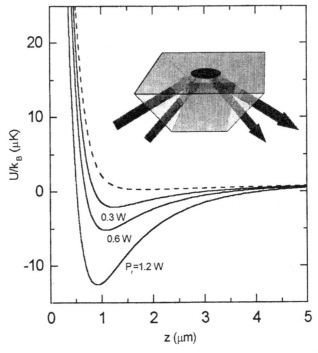

**Fig. 7.9.** Sketch of the dual evanescent wave trap. The light beam making the smaller angle (0.2°) with respect to the critical angle for total internal reflection in the dielectric surface comes from the red-detuned laser. The blue-detuned beam makes an angle of 3.2° above the critical angle and therefore does not penetrate as far in the vertical direction above the prism. The increasing depth of the confining potentials corresponds to increasing power in the red-detuned laser.

trap can attempt evaporative cooling of a sample of $^{133}$Cs pumped to the $F = 3 M = 3$ state. In [162] this group reports temperatures as low as 100 nK and phase space densities near 0.1 in atom samples that have not, as yet, been spin polarized. Spin polarization, which is straightforward to accomplish, will increase the concentration of atoms in one quantum state by as much as a factor of 7. Their results indicate that heating from three-body collisions remains a barrier to the quantum degenerate regime, but the ability to modify the scattering length using a nearby Feshbach resonance should provide the key to effective suppression of the three-body rate. When the phase space density approaches unity the BEC regime is close at hand.

In fact, the Innsbruck group reported achieving the long-sought BEC in $^{133}$Cs using an all-optical trap and some of the techniques already discussed for gaining phase space density [438]. The strategy consists of four essential steps: (1) capture of *many* Cs atoms in a large-volume, shallow potential well using a crossed-beam $CO_2$ laser for realizing an optical dipole-gradient trap. This dipole trap is loaded by adiabatic release from an optical lattice in which the atoms are Raman-side-band cooled [399] to $\sim 1$ μK. The number of atoms loaded into the $CO_2$ trap is about $2 \times 10^7$, all of which are in the ground hyperfine state and spin polarized $F = 3$, $M_F = 3$. Since the optical dipole trap is only about 10 μK

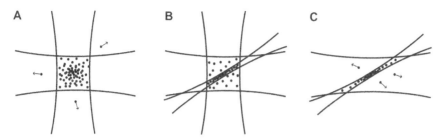

**Fig. 7.10.** (A) Large-volume, crossed-beam $CO_2$ dipole-gradient trap loaded from an adiabatically expanded optical lattice. The depth is about 10 µK. (B) "Dimple" trap, tightly focused ($\simeq$30 µm diameter) into the center of the $CO_2$ trap. (C) Forced evaporative cooling from the atoms in the dimple trap. Forced evaporation at this stage is carried out by ramping down the intensity of the Yb fiber laser, used to make the dimple trap, from 90 mW to a few mW.

deep, a magnetic levitation field must be used to support the atoms against the downward pull of gravity. (2) The first stage of evaporative cooling takes place in the relatively low-density $CO_2$ trap ($\sim 10^{11}$ cm$^{-3}$) by *tuning the scattering length* with a uniform external bias magnetic field to $A_0 \simeq 1200$ a$_0$. In addition to magnetic levitation, Grimm and his research team apply this uniform magnetic field to take advantage of the broad Feshbach resonance around 8 G. Below this field amplitude, the scattering length turns strongly negative (at zero bias field $A_0 = -3000$ a$_0$). At higher field the scattering is positive and increases rapidly. The large amplitude of $A_0$ (at $B = 55$ G, $A_0 = 1000$ a$_0$) increases the two-body elastic collision rate sufficiently to allow evaporative cooling to proceed. (3) The next step is to gain atom density (without heating) by tightly focusing a "dimple" dipole trap from a Yb fiber laser into the middle of the $CO_2$ trap. This technique is similar to that used in the earlier GOST trap experiments. The power in the Yb laser is ramped up to about 90 mW. At the relatively high spatial density achieved with the "dimple" ($\simeq 10^{12}$ cm$^{-3}$), the scattering length has to be magnetically tuned down to $A_0 = 300$ a$_0$ in order to avoid losses from three-body collisions. (4) One of the $CO_2$ lasers is now turned off and forced evaporative cooling out of the dimple trap proceeds by ramping down the power of the Yb fiber laser to a few milliwatts. After the forced evaporative cooling (which takes about 17 s), both $CO_2$ and Yb beams are shut off and the atom cloud is detected within 100 ms of release by absorption imaging. In order to reach phase space density sufficient to trigger the Bose–Einstein condensation, Grimm and coworkers had to run an obstacle course between too high a trapped atom density (giving rise to heating due to three-body collisions) and too low a density (insufficient rate for evaporative cooling). The ability to magnetically tune the scattering length was indispensable. Figures 7.10 and 7.11 show the optical trapping sequence and the resulting trap density profiles after forced evaporation.

## 7.4 Cold molecule formation

Motivation for the study of cold or ultracold molecules derives from two interests: analogous to cold atoms, it would be desirable to confine, manipulate, and transport molecules at the

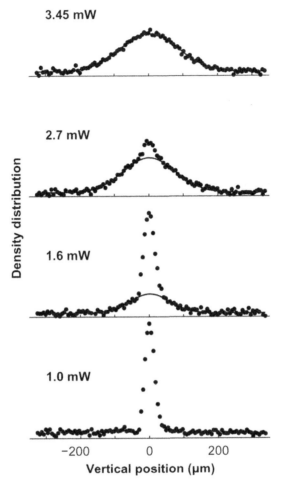

**Fig. 7.11.** Density profiles of the atoms released from the dimple trap after various stages of forced evaporative cooling. The profiles are detected by absorptive imaging after an interval of about 50 ms after the atoms are released from the trap. The onset of BEC is visible at 2.7 mW. Essentially all the atoms are condensed at 1 mW.

individual level in a general effort to control and construct matter microscopically. In this sense it would be worthwhile to efficiently produce cold molecules from cold atoms even if the molecules were produced "incoherently," i.e. with a tight distribution of external and internal observables (preferably cold internally as well as externally), but with no phase relation between the individual state functions of each molecule. Secondly, ultracold molecules in a quantum degenerate gas can be simply regarded as another eigenstate of a BEC. Coupling and population transfer between the atomic form and the molecular form of the BEC would be analogous to familiar coupling between two quantum states of any two-level system. Population transfer or a "Rabi oscillation" of probability from the atomic form to the molecular form by some field coupling holds out the possibility of producing a macroscopic ensemble of molecules phase-coherent in both external and internal (rovibronic) states.

### 7.4.1 Direct methods of cooling or decelerating molecules

One might imagine at first glance that optical cooling should work for molecules as it does for atoms. All one would have to do is find a molecular cycling transition, in which population is stimulated from a lower to a higher-lying state and allowed to spontaneously emit light back to the lower state. The problem is that it is practically impossible to find a unique spontaneous emission pathway back to the initially excited state. Molecules always have manifolds of vibrational and rotational states associated with any electronic transition, and population very rapidly diffuses into rovibronic states not coupled to the exciting light field. Within a few Rabi cycles the population in the optical cooling cycle is effectively pumped into "dark states." Despite this difficulty Bahns et al. [28] proposed a scheme for optical cooling of the $Cs_2$ molecule by using a manifold of optical frequencies derived from stimulated Raman emission of $Cs_2$ molecular beam. This multiple-single-frequency (MSF) laser would sequentially cool rotational, translational, and vibrational states by chirping the "comb" of frequencies in the MSF laser.

Other direct methods of cooling either translational or internal states of molecules include supersonic beams [362], helium cluster cooling (closely related to sympathetic evaporative cooling) [396], buffer-gas loading of paramagnetic molecules into magnetic traps (another sympathetic cooling technique) [115], and field-deceleration of a supersonic beam of molecules [44, 253]. This field-deceleration technique is an interesting combination of internal-state cooling (from the supersonic expansion) and translational deceleration which, although not strictly a cooling mechanism, results in an ensemble of slow moving molecules that can be trapped. Research on cold molecules, broadly interpreted as molecules with a temperature of less than 1 K, has been reviewed by Bahns et al. [27]. This review provides good summaries and discussions of cold molecule research and is a convenient point of entry in the cold molecule literature up to 1999. The article by Bethlem et al. [44] contains a more recent list of citations (up to about mid-year 2002) and a detailed presentation of the Stark decelerator method.

### 7.4.2 Cold molecules from cold atoms: photoassociation

Collisions participate in the production of cold molecules in three different ways: (1) *evaporative cooling* [216] consists essentially of a selective filtering and ejection of a "hot" fraction of molecules from a Maxwell–Boltzman distribution, followed by thermalization of the ensemble remainder by elastic collisions, (2) *sympathetic cooling* [115, 443] in which molecules are cooled by elastic collisions with a reservoir of cold atoms, themselves sometimes actively cooled by light and (3) *photoassociation* in which the center-of-mass kinetic energy of the formed molecule is as translationally cold as its constituent atoms. Here we will concentrate on cold molecule formation by photoassociation since this collisional process has already been treated in Chapter 5, and to date this method produces the translationally coldest molecules.

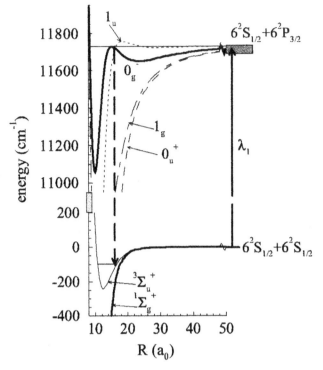

**Fig. 7.12.** Optical transitions and molecular states figuring in the efficient production of $^3\Sigma_u^+$ Cs$_2$ molecules. Note the initial excitation at long range and the favorable Franck–Condon overlap between the inner turning point of the excited $0_g^-$ state and the "ground" triplet state, $^3\Sigma_u^+$ which maximizes the spontaneous emission probability.

### *Molecule stabilization by spontaneous processes*

Although the idea of forming cold molecules by photoassociation had been discussed since the original cold-atom photoassociation proposal [385], interest really crystalized around the topic after Fioretti *et al.* [134] observed copious amounts of Cs$_2$ falling out of their MOT. The unusually efficient production of molecules ($\simeq 10^9$ Cs$_2$ s$^{-1}$) was immediately explained by the fortuitous shape and placement of the relative molecular electronic potential curves. Figure 7.12 shows how the molecules are made. Photoassociation requires a double interaction: initial excitation of the approaching atoms[1] followed by stabilization (usually either by spontaneous or stimulated emission). In the case of Cs$_2$ ground-state triplet formation, the distribution of atom pairs (or pair correlation function) favors initial excitation to a "pure-long-range" excited state at large internuclear separation ($\simeq 100\,a_0$), and the subsequent efficient overlap between the inner turning points of the $0_g^-$ excited state

---

[1] Sometimes a double excitation is needed to get good Franck–Condon overlap with the stabilized molecule. Efficient production of ground singlet state K$_2$ is a case in point.

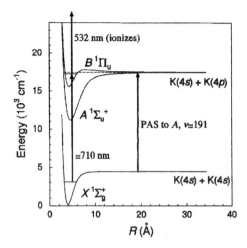

**Fig. 7.13.** Cold-molecule formation of the $K_2$ X $^1\Sigma_g$ ground state. Initial excitation is to a very high-lying vibration level of the A $^1\Sigma_u^+$ state ($^1 0_u^+$ in Hund's case (c) notation). A small fraction ($\simeq 0.15\%$) subsequently emits spontaneously to a few excited vibration levels of the ground-state. The molecules are detected by multiphoton ionization through the B $^1\Pi_u$ intermediate.

and the $^3\Sigma_u^+$ triplet ground state enhances the stabilization step. Soon after this initial report Knize and coworkers reported [378] the observation $Cs_2$ molecules produced in a MOT and trapped in an optical dipole trap.

The Connecticut group headed by P. L. Gould has produced $K_2$ molecules in the X $^1\Sigma_g$ ground state, first by using a two-step process analogous to that for $Cs_2$, and then using an initial double excitation to a molecular Rydberg state followed by spontaneous emission to the ground state. Figure 7.13 shows the "pump and dump" pathway to ground-state $K_2$ molecules. Rather than using a "pure long-range" state as the initial excitation target, they chose to excite high-lying vibration levels of A $^1\Sigma_u$ state. A small fraction of these excited bound-state molecules ($\simeq 0.15\%$) decay to the singlet ground-state. Owing to the offset of the A- and X-state potential curves, however, only highly excited vibrational levels of the X state are populated as shown by the Franck–Condon distribution in Fig. 7.14. In order to achieve better efficiency and populate lower-lying vibrational levels of the ground state the Gould group adapted an idea, proposed earlier by Band and Julienne [33], of a two-step excitation: first to the A state and then from there to a higher-lying Rydberg state. The basic idea is to get better alignment of the potential curves prior to radiative relaxation to the ground state. The scheme adopted is shown in Fig. 7.15. Application of the double excitation results in a much better overlap with low-lying vibration levels of X $^1\Sigma_g$ and more efficient production of molecules. Nikolov *et al.* [306] estimate an improvement of about a factor of 100 in molecule formation rate from the single- to the double-excitation scheme, although it appears that even with this "efficient" scheme the molecule production rate is much below that of $Cs_2$.

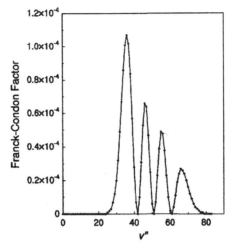

**Fig. 7.14.** Calculated Franck–Condon overlap factors between $v' = 191$ of the A state and a range of $v''$ vibrational levels in the X state. Note that maximum overlap occurs for $v'' = 36$ and that overlap to $v'' = 0$ is negligible.

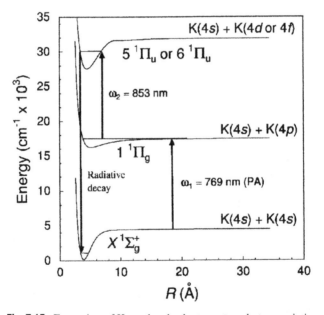

**Fig. 7.15.** Formation of $K_2$ molecules by two-step photoassociation.

An Italian collaboration [140] has observed cold rubidium molecules, both $^{85}Rb_2$ and $^{87}Rb_2$, and measured the molecular spectra. The molecules are produced in a conventional MOT by photoassociation. After spontaneous emission the molecules stabilize in the $a\,^3\Sigma_u^+$ triplet ground state are detected by a pulsed two-photon ionization to $Rb_2^+$. The translational temperature was determined to be near 90 μK.

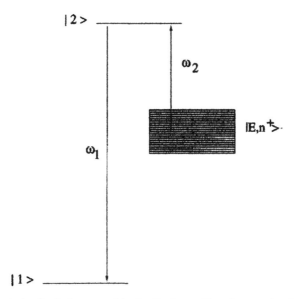

**Fig. 7.16.** The two-photon "radiative recombination" scheme. Note that $\omega_1$ stimulates the population transfer to a specific rovibronic final state.

## *Molecule stabilization by stimulated processes*

At first glance it might seem much more desirable to use stimulated rather than spontaneous emission in the molecule stabilization step since the stimulated process will couple to a specified target internal state and not to a manifold of internal states, self-selected by Franck–Condon overlap. The advantage of spontaneous emission, however is that it is irreversible. Once the molecules form they cannot return to the initial continuum states from whence they came. Stimulated emission, in contrast, is coherent and the rate of coupling forward to the target molecule state will quickly be equalled by the rate of coupling back to the initial continuum states *unless* the back coupling is interrupted in some way, either by irreversible loss out of the target state or by the use of time-dependent coupling fields.

An early systematic scheme for cold molecule formation using stimulated emission to actively stabilize the bound-state molecule was proposed by Vardi *et al.* [411], who suggested that a pair of short, intense *pulses* could be used to direct the population to the lowest vibrational state of the molecular ground electronic state and then terminate before back-coupling could become significant. The basic idea is presented in Fig. 7.16, Vardi *et al.* [411] calculated the population transfer to the molecular bound state with both the "intuitive" pulse sequence, i.e. the first laser pulse (the pump pulse) centered at $\omega_1$ followed by the second pulse (the Stokes pulse) centered at $\omega_2$ and with the "counterintuitive" sequence in which the Stokes pulse is applied first. The "counterintuitive" scheme is reminiscent of STIRAP (stimulated Raman adiabatic passage), a well-known technique [40] for coherently transferring population among bound atomic and molecular states. Application of the Vardi *et al.* scheme to a specific example, cold-molecule formation in Na photoassociation to the ground rovibrational level of the $X\,^1\Sigma_g^+$ Na$_2$ ground electronic state, showed however that

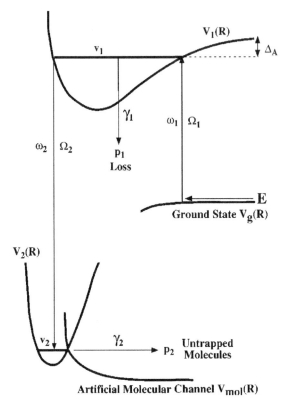

**Fig. 7.17.** Scheme of Julienne *et al.* [200] showing Raman transitions from the entrance channel
continuum state ($V_g(R)$) to the excited intermediate molecular state ($V_1(R)$) and from there to the
final molecular state ($V_2(R)$). The pump laser $\omega_1$ couples $V_g(R)$ to $V_1(R)$ with Rabi frequency $\Omega_1$,
and the Stokes laser $\omega_2$ couples $V_1(R)$ to $V_g(R)$ with Rabi frequency $\Omega_2$. Spontaneous emission loss
rate from the intermediate state is indicated by $\gamma_1$ and the loss rate out of the final target state (rate of
formation of untrapped molecules) is $\gamma_2$.

very large average powers ($\sim 200$ W) were required to transfer half the atom population
to molecules in a reasonable amount of time (a few minutes). The chief difficulty for efficient stimulated transfer to low-lying vibrational states is poor Franck–Condon overlap
between the final target state and the intermediate state (in this case the A $^1\Sigma_u^+$, $v' = 34$,
$J = 1$).

A number of other groups have studied the feasibility of stimulated, two-color photoassociation [86, 200] and, in particular, Julienne *et al.* [200] have proposed a scheme for
producing a molecular form of BEC from the atom condensate through a coherent Raman
process. They begin with a three-state coupling scheme similar to that of Vardi *et al.*, but
their model assumes cw, not pulsed, optical fields; and therefore provides for irreversibility
through a molecular loss channel in the target state with width $\gamma_2$. Figure 7.17 outlines the
scheme. Motivated by [200], Heinzen *et al.* [168] undertook to investigate the possibility of
coherent molecule formation by a stimulated photoassociation in a $^{87}$Rb BEC. Their study
calculated "giant oscillations between an atomic and a molecular condensate," which they

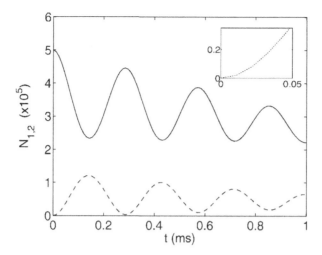

**Fig. 7.18.** Results of model calculations from [168], showing oscillations of population between an atom BEC and the corresponding molecular BEC in which the two are coupled coherently by a Raman pump-stokes transitions in a scheme quite similar to that of Fig. 7.17; except that this model does not take into account loss from the target molecular state.

took to be the signature of a phase-coherent conversion. Furthermore, coherent conversion implies that the conversion rate is density dependent. Figure 7.18 shows an example of these oscillations calculated for reasonable experimental parameters of detuning and laser power. As shown in Fig. 7.19, the University of Texas group headed by D. Heinzen did indeed observe stimulated Raman molecule formation in a $^{87}$Rb BEC [450], but they observed no evidence of oscillations or other molecular BEC signature, and it is presumed that the molecules are not formed in a condensate phase. These results did show, however, the influence of the mean-field interaction on molecule formation. As the BEC density increased the loss-dip due to stimulated molecule formation shifted and increased in width. In a follow-up study, Drummond *et al.* [117] took up again the STIRAP idea for increasing the efficiency of stimulated molecular BEC formation and took into account mean-field collisional dephasing which had been absent in the earlier work [168]. The principle attraction of the stimulated adiabatic passage idea is that population can be transferred from the initial state (atom continuum state in the atom BEC) to the final state (stable bound molecular state in the molecular BEC) without actually populating levels of the excited molecular intermediate and thus suppressing dephasing losses due to spontaneous emission from the intermediate state. As did Vardi *et al.*, however, they concluded that fairly high laser powers ($\simeq 1$ W into a focused beam waist of $10\,\mu$m) were required to obtain efficient conversion ($\geq 90\%$). The reason is that the mean field of the atomic BEC produces dephasing interactions with the atoms undergoing the STIRAP association, and this dephasing time sets a lower limit on the Rabi frequency coupling for the pump and Stokes lasers. In order to effect the STIRAP transfer before destructive collisional dephasing set in, the molecular Rabi frequencies coupling the initial and final states have to greater than about $2 \times 10^7$ s$^{-1}$. The molecular Rabi frequencies are in fact the product of nuclear Franck–Condon overlap factors and the usual

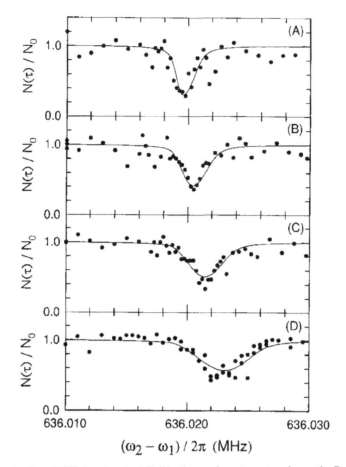

**Fig. 7.19.** Results from [450] showing the "dip" in the condensate centered near the Raman resonance for molecule formation. The resonant loss is presumably due to the formation of molecules. Panels (a)–(d) show a shift and broadening due to increasing density in the BEC. In (a) the density is $N_0 = 0.77 \times 10^{14}$ and increases to $N_0 = 2.60 \times 10^{14}$ in (d). The shift and broadening is evidence of the influence of the mean field interacting with the Raman process.

electronic transition dipole coupling. Poor Franck–Condon overlap requires an electronic Rabi frequency on the order of $10^{11}$ s$^{-1}$, which explains the need for watts of laser power. Nevertheless with judicious choice of laser detunings so as to compensate for level shifts induced by the BEC mean field interactions, nonequal Rabi frequencies for the pump and Stokes pulses, and target vibrational levels Drummond *et al.* [117] were able to predict reasonable conversion efficiencies ($\simeq 50\%$) for the stimulated formation of the molecular condensate using laser powers on the order of tens of milliwatts rather than watts.

### 7.4.3 Molecular BECs from Feshbach resonances

Raman photoassociation is not the only route to a molecular condensate. Stimulated by the experimental observation [361] of the MIT group led by W. Ketterle, that sweeping

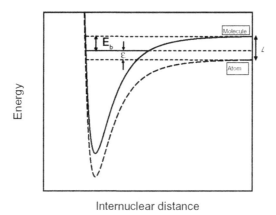

Internuclear distance

**Fig. 7.20.** Schematic potential curves of the model proposed by Timmermans *et al.* [395] showing the entrance channel atomic collision asymptote (dashed curve) and the molecular potential with vibrational level (solid curve). The energy difference between the bound molecular state and the atom entrance channel is $\epsilon$ and can be varied by an external magnetic field. The frequency $\omega$ of the oscillations in Fig. 7.21 is given by $\epsilon/\hbar$.

a magnetic field close to or through a Feshbach resonance produces unexpectedly large inelastic losses, theory [395, 397, 409] suggested that the Feshbach resonance may be coupling atomic and molecular forms of the condensate. The same Feshbach interaction that is used to tune the scattering length can conceivably be used to couple an atomic and energy-degenerate molecular BEC at or near the zero-energy resonance. Furthermore, in an analogy with the STIRAP sweep through the Raman two-photon resonance, a further theoretical study by Mies *et al.* [276] predicted that ramping a magnetic field so as to sweep through a Feshbach resonance may efficiently convert the atom continuum entrance channel to a bound molecule, independently of whether the atoms and molecules are confined in a BEC. As in the stimulated Raman photoassociation, the signature would be a Ramsey-fringe-like oscillation [325] between the atomic and molecular BEC with the frequency of oscillation determined by the energy difference between the asymptotic levels of the atomic and molecular states. The theory group of Verhaar in Eindhoven [409] and Timmmermans *et al.* at the Harvard-Smithsonian Center for Astrophysics [395] developed simple atom–molecule models, and predicted this "signature" oscillation as shown in Figs. 7.20 and 7.21. The JILA group headed by Carl Wieman has reported experimental observation of just such a temporal oscillation in the number of condensate atoms in an $^{85}$Rb BEC. In the experiment, a bias magnetic field is ramped close to a Feschbach resonance for a variable time $t_{evolve}$ during which the presumed two forms of the BEC, atomic and molecular, are coupled. Figure 7.22 shows the time sequencing of the magnetic field, and Fig. 7.23 shows their striking result. Although this experiment does not directly detect the molecules, Donley *et al.* [114] carried out a series of measurements of the oscillation period as a function of magnetic field to test the idea that this period is directly related to the energy difference between the atom and molecule forms. They report that a "sophisticated coupled-channels scattering calculation" [229] provides excellent agreement between observed and calculated oscillatory periods.

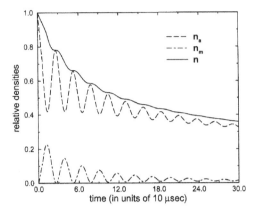

**Fig. 7.21.** Tunneling oscillations between atomic and molecular form of a BEC calculated by [395]. These oscillations resemble those of Fig. 7.18, although in that case the atomic and molecular forms of the BEC are coupled by a Raman optical coupling whereas here the Feshbach resonance is realized through the adjustment of an external magnetic field. The number of atoms $n_a$ and the number of molecules $n_m$ oscillate out of phase.

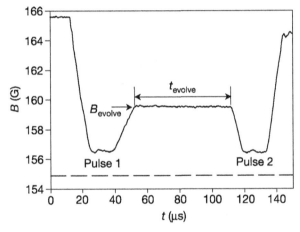

**Fig. 7.22.** Magnetic field pulsing sequence used by Donley *et al.* [114] to induce coupling between atomic and molecular forms of a BEC in $^{85}$Rb. The atom condensate is first brought to a bias field at $\sim 162$ G ($a_s = 10$ $a_0$), then quickly ramped to $\sim 157$ G ($a_s = 2500$ $a_0$) close to the Feshbach resonance indicated by the dashed line. After $\sim 20$ $\mu$s the magnetic field is again raised to $\sim 160$ G ($a_s = 570$ $a_0$) and held there while the coupled system is allowed to evolve for a time $t_{evolve}$.

However, the interpretation of these results appears to be (at the time of writing, October 2002), somewhat unsettled. On the one hand, in a series of theory studies Mackie *et al.* [191, 251, 252] call into question this simple model. In particular, they conclude that *rogue dissociation*, in which a dominant fraction of the formed molecules do *not* dissociate back to the Bose-condensed atom state but to thermal states associated with the confining magnetic trap, damps considerably the coherent atom–molecule BEC oscillations.

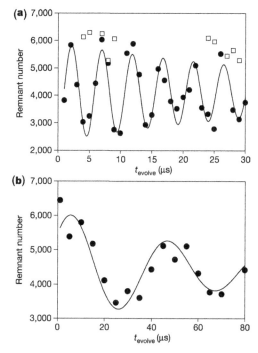

**Fig. 7.23.** Oscillations in the number of condensate atoms after subjecting an atomic $^{85}$Rb BEC to a pair of magnetic field pulses separated by a variable time $t_{evolve}$. Donley *et al.* [114] interpret the periodic loss as being due to coherent coupling to a molecular BEC. Note the difference in period $\omega$ between the top and bottom panels, due to the change in magnetic field and hence the energy difference $\epsilon/\hbar$ as shown in Fig. 7.20.

On the other hand, the work of the theory group led by K. Burnett "strongly supports the view that the [Donley *et al.*] experiments have produced a molecular condensate" [225]. Obviously direct detection of the molecular phase of the BEC would settle the issue.

## 7.5 Collisions and quantum computation

In recent years the close affinity between the mathematics of information science and the physics of quantum mechanics has given birth to a rapidly evolving field of research, quantum information science or quantum computing. The basic motivation and excitement here is the possibility that quantum physics may be regarded not only as a way to characterize the internal, external, and time-evolution of a complete set of commuting observables, but may also be interpreted to describe the information content of physical systems. The classical unit of information is the binary digit or bit. It can be prepared in either one of two distinguishable states, 0 or 1, for example. These states of the bit correspond to classical states of physical systems used to represent them, such as electrical voltage or current or magnetic spin orientation. In quantum mechanics the states of physical systems are thought of as vectors in an orthonormal Hilbert space, the dimensionality of which is prescribed by the complete set of observables. An arbitrary state of a physical system can then be expressed as some linear combination of the basis vectors or "eigenstates" of the system. The basic idea of "quantum information" is that its unit, the quantum bit or qubit may also be described in a Hilbert space of two dimensions corresponding to the two observable values of the physical system expressing it, Thus, for example,

$$|\psi\rangle = \frac{1}{\sqrt{2}} (|0\rangle + |1\rangle)$$

where $|\psi\rangle$ is the qubit. Numbers can be represented and operations on them implemented using qubits rather than classical bits. One can form a $2^n$-dimensional space in which the basis vectors are orthogonal and consist of linear combinations of $n$ qubits. For example, one could represent quantum-information "words" of two-qubit length in a four-dimensional space. A general state in this space would have the form

$$|\psi\rangle = a|00\rangle + b|01\rangle + c|10\rangle + d|11\rangle,$$

where the basis vectors are formed from the tensor product of two qubits and the coefficients are subject to a normalization condition. The general state is said to be *entangled* when it cannot be represented as a simple product state of the basis vectors. A quantum *gate* is a unitary operator which transforms a qubit, or a register of $n$ qubits, from one state vector into another. Here are some examples of one-qubit gates taken from Stean [359],

identity gate        $I \equiv |0\rangle\langle 0| + |1\rangle\langle 1|$

NOT gate        $X \equiv |0\rangle\langle 1| + |1\rangle\langle 0|$

Hadamard gate        $H \equiv \frac{1}{\sqrt{2}} [(|0\rangle + |1\rangle)\langle 0| + (|0\rangle - |1\rangle)\langle 1|]$.

The two-qubit networks have particular relevance to binary collisions between atoms in well-defined internal and external states. Here are is an example of the result of a gate operation on a network of two qubits,

CNOT gate

$|00\rangle$        $\rightarrow$        $|00\rangle$

$|01\rangle$        $\rightarrow$        $|01\rangle$

$|10\rangle$        $\rightarrow$        $|11\rangle$

$|11\rangle$        $\rightarrow$        $|10\rangle$.

The term CNOT means "controlled NOT gate." The first (left) bit acts as the control bit and the second (right) bit is the target bit. From the table it is easy to see that the NOT operation is controlled by the first bit. When that bit is 0 the NOT operation is turned off, and the target bit is unchanged. When the control bit is switched to 1 the NOT operation acts on the target bit. Here is another example of a two-qubit operation.

CPHASE gate

$|00\rangle$　　　$\rightarrow$　　　$|00\rangle$

$|01\rangle$　　　$\rightarrow$　　　$|01\rangle$

$|10\rangle$　　　$\rightarrow$　　　$|10\rangle$

$|11\rangle$　　　$\rightarrow$　　　$-|11\rangle$.

The CPHASE gate means "controlled PHASE gate," and it is clear that when the control bit is turned on and the target bit is also in the high state, the only effect is to multiply the two-qubit state by a phase $e^{i\pi}$. One could easily imagine, for example, a two-level atom with "spin up" and "spin down" basis states acting as the unit qubit. Collisional interaction between the two atoms would entangle the two qubits. If each atom were spatially confined and could be "collided" in a controlled way, then two-qubit quantum gates might be implemented. The precise definition of internal and external states and the controlled interaction immediately suggest ultracold atoms in adjustable periodic potentials such as optical lattices.

The physical requirements for the realization of a quantum computer has been discussed by DiVencenzo [110] who has identified five essential characteristics of any physical system implementing such a device. These characteristics are: (1) scalability, (2) initialization of all qubits to a reference state, (3) long decoherence times, (4) fast gate times, and (5) the ability to measure individual qubits. Cold atoms in lattices, assuming the lattice can be filled, constitute a naturally scalable system, and initialization of the internal and external states would not be difficult to arrange. Thus characteristics (1) and (2) are (in principle) easily fulfilled. Criteria (3) and (4) form a pair because "long" coherence time means long compared to the gate operation. Strongly interacting particles would certainly yield fast gate times, but it is not easy to find a system in which the two qubits are strongly interacting with each other but weakly interacting with the surroundings (long coherence time). Criterion (5), measuring the final state of the qubit, can be accomplished (again in principle) by state-selective, high-resolution laser excitation and should therefore pose no fundamental problem.

Several proposals are extant. For example, Jaksch et al. [188] have proposed that atoms trapped in optical lattices be brought together in a controlled way. The energy or phase shift of this controlled collisional interaction is directly proportional to the s-wave scattering length, $a_s$. Figure 7.24 describes the scheme to implement a CPHASE gate. The basic idea is to trap atoms with internal states $|a\rangle$ and $|b\rangle$ in intercalated optical potentials. Then, by changing the relative angle of polarization in the standing waves constituting the optical lattice, one can systematically bring the atoms in state $|a\rangle$ physically close to atoms in state $|b\rangle$. The collisional interaction energy is

$$\Delta E(t) = \frac{4\pi a_s \hbar^2}{m} [\langle a(x,t)|a(x,t)\rangle \langle b(x,t)|b(x,t)\rangle]$$

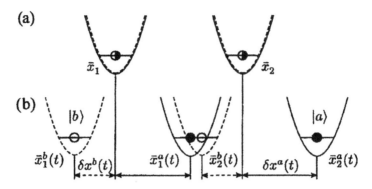

**Fig. 7.24.** atoms in internal states $|a\rangle$ and $|b\rangle$ are brought together spatially in a controlled way. The atom states are allowed to interact, and then the atoms are again spatially separated. The atom states are now entangled as indicated in (a).

and the collisional phase accumulates over the time of the interaction,

$$\phi^{ab} = \frac{1}{\hbar} \int_{-\tau}^{+\tau} dt\, \Delta E(t).$$

The action of the gate on the two-qubit network is then

$$|a\rangle|a\rangle \rightarrow |a\rangle|a\rangle$$
$$|a\rangle|b\rangle \rightarrow e^{i\phi_{ab}}|a\rangle|b\rangle$$
$$|b\rangle|a\rangle \rightarrow |b\rangle|a\rangle$$
$$|b\rangle|b\rangle \rightarrow |b\rangle|b\rangle$$

which is clearly a CPHASE gate with the left-hand bit acting as the control bit ($|a\rangle$ is "high") and the right-hand bit as the target. Figure 7.25 shows a 3-D picture of the optical lattice and how atoms in state $|a\rangle$ (black balls) can be brought into collisional interaction with atoms in state $|b\rangle$ (white balls) by changing the polarization angle $\theta$ indicated on the left-side side of the figure.

Another example is the suggestion proposed by Deutsch and Brennan [109] in which the entanglement proceeds by resonant dipole–dipole interactions thus speeding up the gate operation time. The atoms are confined by an optical grating similar to that envisaged by Jaksch *et al.* [188]. In addition to proposing a CPHASE gate, similar to [188], they also develop a gate operating on atom qubits not labeled by the atom internal states but by their vibrational quantum numbers in a 3-D harmonic well. The advantage of this approach is that the collisional interaction is not diagonal in this basis and the action of the gate is to effect real population changes, not just phase shifts. The challenge here will be to prepare all the atoms in the correct trap vibrational states.

An obvious question is: are there cases where there is a significant advantage in constructing information with qubits rather than with classical bits? The answer is yes. P. W. Shor [352] in 1994 described an algorithm for factoring large numbers that would be exponentially faster on a quantum computer than on a classical machine. Large-number factorization

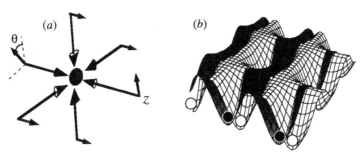

**Fig. 7.25.** Atoms with two different internal states (black balls and white balls) confined in two intercalated optical lattices. The distance between the black-ball lattice and the white-ball lattice is controlled by the "lin $\theta$ lin" polarization arrangement of the standing wave as indicated in (a).

is an essential task for analyzing encryption keys that permit the reading of enciphered communications. Shor's algorithm therefore has important implications for the maintenance of secure communications. The qubits intrinsic to quantum computing are also important to related areas of quantum information such as data compression and cryptography [359].

## 7.6 Future directions

At the time of writing (January 2003) ultracold collisions continue to attract the attention of a significant number of researchers working in various fields of atom manipulation. In this section we briefly sketch some topics that appear particularly promising.

Collisions in a light field present several intriguing areas of investigation. For example, proposals for cooling and trapping molecules have recently appeared [28, 33, 115] that would extend a Doppler-free precision spectroscopy to species not susceptible to direct optical cooling and may provide the starting point for building trapped clusters. Foretti *et al.* [134] have reported a cold molecule by photoassociation and described the observation of translationally cold $Cs_2$ molecules falling out of a MOT. Julienne *et al.* [200] have proposed that cold molecules can be efficiently made by a stimulated Raman process in a Bose–Einstein condensate. The prospects for producing cold molecules raise numerous issues about collisions between two cold molecules, or a cold molecule and a cold atom. Multiple-color experiments, in which double-resonance optical excitation accesses "pure long-range" intermediate states, have begun the spectroscopic characterization of heretofore inaccessible highly excited and cold molecular Rydberg states [429]. Two recent reports describe the production of a "frozen Rydberg gas" after excitation of atomic Rydberg states in cold Cs [296] and Rb [16]. These experiments reveal novel many-body phenomena in a dilute gas. Collisions within and between highly collimated, dense, bright atomic beams reveal alignment and orientation features of ultracold collisions that are usually obscured by spatial averaging in trap collisions. A few steps have already been taken in this direction [384, 405, 437, 439] but these only indicate the potential for future studies. For example, atomic beam collisions will lead to a simplification of the analysis of photoassociation spectra through alignment and spin polarization [146]. Spin orientation effects suppressing

Penning ionization in He metastable and other rare-gas metastable collisions can be also directly probed in beams. Doery *et al.* [112] have calculated the long-range potentials of noble gas dimers, showing the existence of pure long-range states that could be probed by photoassociation spectroscopy. Tiesinga *et al.* [393] have developed computational methods which treat nuclear and electron spin explicitly and permit accurate calculations of molecular hyperfine structure in photoassociation spectra. Atom association between mixed species confined in "dual MOTs" will extend precision molecular spectroscopy from homonuclear to heteronuclear diatomic species. Initial studies with dual MOTs are already under active investigation in Sao Carlos and Rochester (Santos *et al.* [339], Optical Society of America [308]). Development of convenient laser sources in the blue and near-ultraviolet regions of the spectrum will more readily permit cooling and trapping of Group IIA atoms such as Mg, Ca, Sr, and Ba. The common isotopes of the alkaline earths have no nuclear spin and therefore exhibit no hyperfine interaction. Photoassociation spectra and trap-loss spectra should therefore be interpretable with a rigorous, unifying, quantitative theory, including effects of strong light fields and spontaneous emission. The study of optical shielding and suppression in nuclear-spin-free systems will also permit a rigorous comparison of experiment and theory. Up to the present most cold and ultracold collision measurements have been carried out in the frequency domain. Precision molecular spectroscopy has been the dominant influence. Some collision processes, however, lend themselves to the time domain; and the observation of ultracold collisions in real time [149, 307] and with wave packets [411] may open a fruitful new avenue for studies of dynamics and molecule formation complementary to high-precision frequency spectroscopy.

Ground-state collisions will continue to play an invaluable role as a probe of dilute, gas-phase quantum statistical condensates. Output coupling of confined BECs for the purpose of realizing beams of atoms in a single quantum state, "atom lasers" [272] will need to address the issue of collision-limited coherence times. Details of scattering lengths between mixed species will determine the feasibility of sympathetic cooling [299] as a useful technique to extend BEC to a greater range atomic or molecular systems. For example, [58, 59] discuss how $^{87}$Rb can be used for sympathetic cooling of the $^{85}$Rb isotope, which is difficult to cool directly because of a small elastic collision cross section in the relevant temperature range. Control of the sign and magnitude of the scattering length by magnetic [58, 59, 416], rf, or optical field manipulation [128] will be the focus of intensive efforts because a large positive scattering length provides the only demonstrated route to BEC via evaporative cooling.

Although we have discussed the principal future directions of which we are aware, the field of cold and ultracold collisions is presently germinating new ideas at such a rate that any assessment purporting to characterize the overall direction or emphasis even in the near term would be foolhardy. The only certitude one can express without fear of contradiction will be the need to update and supersede this review with another one in the near future.

# References

[1] Abraham E. R. I., W. I. McAlexander, J. M. Gerton, and R. G. Hulet. Singlet s-wave scattering lengths of $^6$Li and $^7$Li. *Phys. Rev. A*, 53: R3713–R3715, 1996.

[2] Abraham E. R. I., W. I. McAlexander, J. M. Gerton, and R. G. Hulet. Triplet s-wave resonance in $^6$Li collisions and scattering lengths of $^6$Li and $^7$Li. *Phys. Rev. A*, 55: R3299–3302, 1997.

[3] Abraham E. R. I., W. I. McAlexander, C. A. Sackett, and R. G. Hulet. Spectroscopic determination of the s-wave scattering length of lithium. *Phys. Rev. Lett.*, 74: 1315–1318, 1995.

[4] Abraham E. R. I., W. I. McAlexander, H. T. C. Stoof, and R. G. Hulet. Hyperfine structure in photoassociative spectra of $^6$Li$_2$ and $^7$Li$_2$. *Phys. Rev. A*, 53: 3092–3097, 1996.

[5] Abraham E. R. I., W. M. Ritchie, W. I. McAlexander, and R. G. Hulet. Photoassociative spectroscopy of long-range states of ultracold $^6$Li$_2$ and $^7$Li$_2$. *J. Chem. Phys.*, 103: 7773–7778, 1995.

[6] Adams C. S., O. Carnal, and J. Mlynek. Atom interferometery. *Adv. At. Mol. Opt. Phys.*, 34: 1–33, 1994.

[7] Adams C. S. and E. Riis. Laser cooling and trapping of neutral atoms. *Prog. Quantum Electron.*, 21: 1–79, 1997.

[8] Adams C. S., M. Sigel, and J. Mlynek. Atom optics. *Phys. Rep.*, 240: 143–210, 1994.

[9] Agosta Ch., I. F. Silvera, H. T. C. Stoof, and B. J. Verhaar. Trapping of neutral atoms with resonant microwave radiation. *Phys. Rev. Lett.*, 62: 2361–2364, 1989.

[10] Ahn R. M. C., J. P. H. W. van den Eijnde, and B. J. Verhaar. Calculation of nuclear spin relaxation rate for spin-polarized atomic hydrogen. *Phys. Rev. B*, 27: 5424–5432, 1983.

[11] Allison A. and A. Dalgarno. Spin change in collisions of hydrogen atoms. *Astrophys. J.*, 158: 423–425, 1969.

[12] Allison. A. C. Spin-change frequency shifts in H–H collisions. *Phys. Rev. A*, 5: 2695–2696, 1972.

[13] Amelink A., K. M. Jones, P. D. Lett, P. van der Straten, and H. G. M. Heideman. Single-color photoassociative ionization of ultracold sodium: the region from 0 to −5 GHz. *Phys. Rev. A*, 62: 013408-1–013408-8, 2000.

[14] Amelink A., K. M. Jones, P. D. Lett, P. van der Straten, and H.G.M. Heideman. Spectroscopy of autoionizing doubly excited states in ultracold Na$_2$ molecules produced by photoassociation. *Phys. Rev. A*, 61: 042707-1–042707-9, 2000.

[15] Anderson M. H., J. R. Ensher, M. R. Matthews, C. E. Wieman, and E. A. Cornell. Observation of Bose–Einstein condensation a dilute atomic vapor. *Science*, 269: 198–201, 1995.

[16] Anderson W. R., J. R. Veal, and T. F. Gallagher. Resonant dipole–dipole energy transfer in a nearly frozen Rydberg gas. *Phys. Rev. Lett.*, 80: 249–252, 1998.

[17] Andrews M. R., C. G. Townsend, H.-J. Miesner, D. S. Durfee, D. M. Kurn, and W. Ketterle. Observation of interference between two Bose–Einstein condensates. *Science*, 275: 637–641, 1997.

[18] Anglin J. R. and W. Ketterle. Bose–Einstein condensation of atomic gases. *Nature*, 416: 211–218, 2002.

[19] Arndt M., M. Ben Dahan, D. Guery-Odelin, M. W. Renolds, and J. Dalibard. Observation of a zero-energy resonance in Cs–Cs collisions. *Phys. Rev. Lett.*, 79: 625–628, 1997.

[20] Ashkin A. Acceleration and trapping of particles by radiation pressure. *Phys. Rev. Lett.*, 24: 156–159, 1970.

[21] Ashkin A. Trapping of atoms by resonance radiation. *Phys. Rev. Lett.*, 40: 729–732, 1978.

[22] Ashkin A. and J. P. Gordon. Stability of radiation-pressure particle traps: an optical Earnshaw theorem. *Opt. Lett.*, 8: 511–513, 1983.

[23] Bagnato V. S., G. P. Lafyatis, A. C. Martin, E. L. Raab, R. Ahmad-Bitar, and D. E. Pritchard. Continuous stopping and trapping of neutral atoms. *Phys. Rev. Lett.*, 58: 2194–2197, 1987.

[24] Bagnato V. S., L. Marcassa, C.-C. Tsao, Y. Wang, and J. Weiner. Two-color spectroscopy of colliding ultracold atoms. *Phys. Rev. Lett.*, 70: 3225–3228, 1993.

[25] Bagnato V. S., L. Marcassa, Y. Wang, J. Weiner, P. S. Julienne, and Y. B. Band. Ultracold photoassociative ionization collisions in a magneto-optical trap: The optical-field-intensity dependence in a radiatively dissipative environment. *Phys. Rev. A*, 48: R2523–R2526, 1993.

[26] Bagnato V. S., J. Weiner, P. S. Julienne, and C. J. Williams. Long-range molecular states and ultracold photoassociative ionization collisions. *Laser Phys.*, 4: 1062–1065, 1994.

[27] Bahns J. T., P. L. Gould, and W. C. Stwalley. Formation of cold ($T \leq 1$ K) molecules. *Adv. At. Mol. Opt. Phys.*, 42: 171–224, 2000.

[28] Bahns J. T., W. C. Stwalley, and P. L. Gould. Laser cooling of molecules: a sequential scheme for rotation, translation, and vibration. *J. Chem. Phys.*, 104: 9689–9697, 1996.

[29] Balakrishnan N., R. C. Forrey, and A. Dalgarno. Threshold phenomena in ultracold atom–molecule collisions. *Chem. Phys. Lett.*, 280: 1–4, 1997.

[30] Balakrishnan N., V. Karchenko, R. C. Forrey, and A. Dalgarno. Complex scattering lengths in multichannel atom–molecule collision. *Chem. Phys. Lett.*, 280: 5–8, 1997.

[31] Bali S., D. Hoffmann, and T. Walker. Novel intensity dependence of ultracold collisions involving repulsive states. *Europhys. Lett.*, 27: 273–277, 1994.

[32] Band Y. B. and P. S. Julienne. Optical-Bloch-equation method for cold-atom collisions: Cs loss from optical traps. *Phys. Rev. A*, 46: 330–343, 1992.

[33] Band Y. B. and P. S. Julienne. Ultracold molecule production by laser-cooled atom photoassociation. *Phys. Rev. A*, 51: R4317–R4320, 1995.

[34] Band Y. B. and I. Tuvi. Reduced optical shielding of collisional loss for laser-cooled atoms. *Phys. Rev. A*, 51: R4329–R4332, 1995.

[35] Band Y. B. , I. Tuvi, K.-A. Suominen, K. Burnett, and P. S. Julienne. Loss from magneto-optical traps in strong laser fields. *Phys. Rev. A*, 50: R2826–R2829, 1994.

[36] Bardou F., O. Emile, J.-M. Courty, C. I. Westbrook, and A. Aspect. Magneto-optical trapping of metastable helium: collisions in the presence of resonant light. *Europhys. Lett.*, 20: 681–686, 1992.

[37] Barrett M. D., J. A. Sauer, and M. S. Chapman. All-optical formation of an atomic Bose–Einstein condensate. *Phys. Rev. Lett.*, 87: 010404-1–010404-4, 2001.

[38] Bayliss W. E., J. Pascale, and F. Rossi. Polarization and electronic excitation in nonreactive collisions: basic formulation for quantum calculations of collisions between $^2$P-state alkali-metal atoms and $H_2$ or $D_2$. *Phys. Rev. A*, 36: 4212–4218, 1987.

[39] Bergeman T., G. Erez, and H. Metcalf. Magnetostatic trapping fields for neutral atoms. *Phys. Rev. A*, 35: 1535–1546, 1987.

[40] Bergmann K., H. Theuer, and B. W. Shore. Coherent population transfer among quantum states of atoms and molecules. *Rev. Mod. Phys.*, 70: 1003–1025, 1998.

[41] Bergmann T. Hartree–Fock calculations of Bose–Einstein condensation of $^7$Li atoms in a harmonic trap for $T > 0$. *Phys. Rev. A*, 55: 3658–3669, 1997.

[42] Berlinsky A. J. and B. Shizgal. Spin-exchange scattering cross sections for hydrogen atoms at low temperatures. *Can. J. Phys.*, 58: 881–885, 1980.

[43] Bethe H. Theory of disintegration of nuclei by neutrons. *Phys. Rev.*, 47: 747–759, 1935.

[44] Bethlem H. L., F. M. H. Crompvoets, R. T. Jongma, S. Y. T. van de Meerakker, and G. Meijer. Deceleration and trapping of ammonia using time-varying electric fields. *Phys. Rev. A*, 65: 053416-1–053416-20, 2002.

[45] Blangé J. J., J. M. Zijlstra, A. Amelink, X. Urbain, H. Rudolph, P. van der Straten, and H. C. W. Beijerinck. Vibrational state distribution of $Na_2^+$ ion created in ultracold collisions. *Phys. Rev. Lett.*, 78: 3089–3092, 1997.

[46] Boesten H. M. J. M., C. C. Tsai, B. J. Verhaar, and D. J. Heinzen. Observation of a shape resonance in cold atom scattering by pulsed phtoassociation. *Phys. Rev. Lett.*, 77: 5194–5197, 1996.

[47] Boesten H. M. J. M. and B. J. Verhaar. Simple quantum-mechanical picture of cold optical collisions. *Phys. Rev. A*, 49: 4240, 1994.

[48] Boesten H. M. J. M., B. J. Verhaar, and E. Tiesinga. Quantum suppression of collisional loss rates in optical traps. *Phys. Rev. A*, 48: 1428–1433, 1993.

[49] Boesten H. M. J. M., J. M. Vogels, J. G. C. Tempelaars, and B. J. Verhaar. Properties of cold collisions of $^{39}$K atoms and of $^{41}$K atoms in relation to Bose–Einstein condensation. *Phys. Rev. A*, 54: R3726–R3729, 1996.

[50] Bohn J. and P. S. Julienne. Prospects for influencing scattering lengths with far-off-resonant light. *Phys. Rev. A*, 56: 1486–1491, 1997.

[51] Bohn J. L. and P. S. Julienne. Semianalytic treatment of two-color photoassociation spectroscopy and control of cold atoms. *Phys. Rev. A*, 54: R4637–R4640, 1996.

[52] Boisseau C., E. Audouard, and J. Vigué. Quantization of the highest levels in a molecular potential. *Europhys. Lett.*, 41: 349–354, 1998.

[53] Boisseau C. and J. Vigué. Laser-dressed molecular interactions at long range. *Opt. Commun.*, 127: 251–256, 1996.

[54] Bolda E. L., E. Tiesinga, and P. S. Julienne. Effective-scattering-length model of ultracold atomic collisions and Feshbach resonances in tight harmonic traps. *Phys. Rev. A*, 66: 013403-1–014403-7, 2002.

[55] Bonanno R., J. Boulmer, and J. Weiner. Determination of the absolute rate constant for associative ionization in crossed-beam collisions between Na $3^2P_{3/2}$ atoms. *Phys. Rev. A*, 28: 604–608, 1983.

[56] Bradley C. C., A. Sackett, and R. G. Hulet. Bose–Einstein condensation of lithium: observation of limited condensate number. *Phys. Rev. Lett.*, 78: 985–988, 1997.

[57] Bradley C. C., C. A. Sackett, J. J. Tollett, and R. G. Hulet. Evidence of Bose–Einstein condensation in an atomic gas with attractive interactions. *Phys. Rev. Lett.*, 75: 1687–1690, 1995.

[58] Burke J., J. L. Bohn, B. D. Esry, and C. H. Greene. Prospects for mixed-isotope Bose–Einstein condensation in rubidium. *Phys. Rev. Lett.*, 80: 2097–2100, 1998.

[59] Burke J. P., J. L. Bohn, B. D. Esry, and Chris H. Greene. Impact of the $^{87}$Rb singlet scattering length on suppressing inelastic collisions. *Phys. Rev. A*, 55: R2511–R2514, 1997.

[60] Burnett K. Bose–Einstein condensation with evaporatively cooled atoms. *Contemp. Phys.*, 37: 1–14, 1996.

[61] Burnett K., P. S. Julienne, P. D. Lett, E. Tiesinga, and C. Williams. Quantum encounters of the cold kind. *Nature*, 416: 225–232, 2002.

[62] Burnett K., P. S. Julienne, and K.-A. Suominen. Laser driven collisions between atoms in a Bose–Einstein condensed gas. *Phys. Rev. Lett.*, 77: 1416–1419, 1996.

[63] Burt E. A., R. W. Ghrist, C. J. Myatt, M. J. Holland, E. A. Cornell, and C. E. Wieman. Coherence, correlations and collisions: what one learns about Bose–Einstien condensates from their decay. *Phys. Rev. Lett.*, 79: 337–340, 1997.

[64] Bussery B. and M. Aubert-Frecon. Multipolar long-range electrostatic, dispersion, and induction energy terms for the interactions between two identical alkali atoms Li, Na, K, Rb, and Cs in various electronic states. *J. Chem. Phys.*, 82: 3224–3234, 1985.

[65] Bussery B. and M. Aubert-Frecon. Potential energy curves and vibration–rotation energies for the two purely long-range bound states $1_u$ and $0_g^-$ of the alkali dimers $M_2$ dissociating to $M(ns\,^2S_{1/2})+M(np\,^2P_{3/2})$ with M = Na, K, Rb, and Cs. *J. Mol. Spectrosc.*, 113: 21–27, 1985.

[66] Cable A., M. Prentiss, and N. P. Bigelow. Observation of sodium atoms in a magnetic molasses trap loaded by a continuous uncooled source. *Opt. Lett.*, 15: 507–509, 1990.

[67] Calarco T., E. A. Hinds, D. Jaksch, J. Schmiedmayer, J. I. Cirac, and P. Zoller. Quantum gates with neutral atoms: controlling collisional interaction in time-dependent traps. *Phys. Rev. A*, 61: 022304-1–02304-11, 2000.

[68] Camposeo A., A. Piobini, F. Cervelli, F. Tantussi, F. Fuso, and E. Arimondo. A cold cesium atomic beam produced out of a pyramidal funnel. *Opt. Commun.*, 200: 231–239, 2001.

[69] Casimir H. B. G. and D. Polder. The influence of retardation on the London–van der Waals forces. *Phys. Rev.*, 73: 360–372, 1948.

[70] Castin Y. and R. Dum. Bose–Einstein condensation in time-dependent traps. *Phys. Rev. Lett.*, 77: 5315–5319, 1996.

[71] Castin Y., K. Mølmer, and J. Dalibard. Monte Carlo wave-function method in quantum optics. *J. Opt. Soc. Am. B*, 10: 524–538, 1993.

[72] Cheng C., V. Vuletić, A. J. Kerman, and S. Chu. High resolution Feshbach spectroscopy of cesium. *Phys. Rev. Lett.*, 85: 2717–2720, 2000.

[73] Chu S., J. E. Bjorkholm, A. Ashkin, and A. Cable. Experimental observation of optically trapped atoms. *Phys. Rev. Lett.*, 57: 314–317, 1986.

[74] Chu S., L. Hollberg, J. Bjorkholm, A. Cable, and A. Ashkin. Three-dimensional viscous confinement and cooling of atoms by resonance radiation pressure. *Phys. Rev. Lett.*, 55: 48–51, 1985.

[75] Clairon A., C. Salomon, S. Guellati, and W. D. Phillips. Ramsey resonance in a Zacharias fountain. *Europhys. Lett.*, 16: 165–170, 1991.

[76] Cline R. A., J. D. Miller, and D. J. Heinzen. Photoassociation spectrum of ultracold Rb atoms. *Phys. Rev. Lett.*, 71: 2204–2207, 1993.

[77] Cline R. A., J. D. Miller, and D. J. Heinzen. Study of $Rb_2$ long-range states by high-resolution photoassociation spectroscopy. *Phys. Rev. Lett.*, 73: 632–635, 1994.

[78] Cline R. A., J. D. Miller, and D. J. Heinzen. Study of $Rb_2$ long-range states by high-resolution photoassociation spectroscopy. *Phys. Rev. Lett.*, 73: 2636(E), 1994.

[79] Cline R. W., T. J. Greytak, and D. Kleppner. Nuclear polarization of spin-polarized atomic hydrogen. *Phys. Rev. Lett.*, 47: 1195–1198, 1981.

[80] Cohen-Tannoudji C., J. Dupont-Roc, and G. Grynberg. *Atom–Photon Interactions: Basic Processes and Applications*. Wiley, New York, 1992.

[81] Cook R. J. Atomic notion in resonant radiation: an application of Earnshaw's theorem. *Phys. Rev. A*, 20: 224–228, 1979.

[82] Cook R. J. Theory of resonant-radiation pressure. *Phys. Rev. A*, 22: 1078–1098, 1980.

[83] Cornell E. Very cold indeed: the nanokelvin physics of Bose–Einstein condensation. *J. Res. Natl. Inst. Stand. Technol.*, 101: 419–434, 1996. and other articles in this special issue on Bose–Einstein condensation.

[84] Cornish S. L., N. R. Claussen, J. L. Roberts, E. A Cornell, and C. E. Wieman. Stable $^{85}$Rb Bose–Einstein condensates with widely tunable interactions. *Phys. Rev. Lett.*, 85: 1795–1798, 2000.

[85] Côté R. and A. Dalgarno. Elastic scattering of two Na atoms. *Phys. Rev. A*, 50: 4827–4835, 1994.

[86] Côté R. and A. Dalgarno. Mechanism for the production of vibrationally excited ultracold molecules of $^7$Li$_2$. *Chem. Phys. Lett.*, 279: 50–54, 1997.

[87] Côté R., A. Dalgarno, and M. J. Jamieson. Elastic scattering of two $^7$Li atoms. *Phys. Rev. A*, 50: 399–404, 1994.

[88] Côté R., A. Dalgarno, Y. Sun, and R. G. Hulet. Photoabsorption by ultracold atoms and the scattering length. *Phys. Rev. Lett.*, 74: 3581–3583, 1995.

[89] Côté R., A. Dalgarno, H. Wang, and W. C. Stwalley. Potassium scattering lengths and prospects for Bose–Einstein condensation and sympathetic cooling. *Phys. Rev. A*, 57: R4118–R4121, 1998.

[90] Côté R., E. J. Heller, and A. Dalgarno. Quantum suppression of cold atom collisions. *Phys. Rev. A*, 53: 234–241, 1996.

[91] Côté R., M. J. Jamieson, Z.-C. Yan, N. Geum, G.-H. Jeung, and A. Dalgarno. Enhanced cooling of hydrogen atoms by lithium atoms. *Phys. Rev. Lett.*, 84: 2806–2808, 2000.

[92] Courteille Ph., R. S. Freeland, D. J. Heinzen, F. A. van Abeelen, and B. J. Verhaar. Observation of a Feshbach resonance in cold atom scattering. *Phys. Rev. Lett.*, 81: 69–72, 1998.

[93] Crampton S. B., W. D. Phillips, and D. Kleppner. Proposed low temperature hydrogen maser. *Bull. Am. Phys. Soc.*, 23: 86, 1978.

[94] Crampton S. B. and H. T. M. Wang. Duration of hydrogen atom spin-exchange collisions. *Phys. Rev. A*, 12: 1305–1312, 1975.

[95] Dalfovo F., S. Giorgini, L. P. Pitaevskii, and S. Stringari. Theory of Bose–Einstein condensation in trapped gases. *Rev. Mod. Phys.*, 71: 463–512, 1999.

[96] Dalgarno A. Spin-change cross sections. *Proc. R. Soc. London A*, 262: 132–135, 1961.

[97] Dalibard J. Laser cooling of an optically thick gas: the simplest radiation pressure trap? *Opt. Commun.*, 68: 203–208, 1988.

[98] Dalibard J. and C. Cohen-Tannoudji. Dressed-atom approach to atomic motion in lasr light: the dipole force revisited. *J. Opt. Soc Am. B*, 2: 1701–1720, 1985.

[99] Dalibard J. and C. Cohen-Tannoudji. Laser cooling below the Doppler limit by polarization gradients: simple theoretical models. *J. Opt. Soc. Am B*, 6: 2023–2045, 1989.

[100] Dasheveskaya E. I. Effect of short-range forces and of twisting on intramultiplet mixing in collision of alkali-metal atoms. *Opt. Spectrosc.*, 46: 236–240, 1979.

[101] Dasheveskaya E. I., A. I. Voronin, and E. E. Nikitin. Theory of excitation trasfer between alklai atoms, I. Identical partners. *Can. J. Phys.*, 47: 1237–1248, 1969.

[102] Davis K. B., M. Mewes, M. A. Joffe, M. R. Andrews, and W. Ketterle. Evaporative cooling of sodium atoms. *Phys. Rev. Lett.*, 74: 5202–5205, 1995.

[103] Davis K. B., M.-O. Mewes, M. R. Andrews, N. J. van Druten, D. S. Durfee, D. M. Kurn, and W. Ketterle. Bose–Einstein condensation in a gas of soidum atoms. *Phys. Rev. Lett.*, 75: 3969–3973, 1995.

[104] De Goey L. P. H., T. H. M. Van der Berg, N. Mulders, H. T. C. Stoof, and B. J. Verhaar. Three-body recombination in spin-polarized atomic hydrogen. *Phys. Rev. B*, 34: 6183–6191, 1986.

[105] De Goey L. P. H., H. T. C. Stoof, B. J. Verhaar, and W. Glockle. Role of three-body correlations in recombination of spin-polarized atomic hydrogen. *Phys. Rev. B*, 38: 646–658, 1988.

[106] DeGraffenreid W., J. Ramirez-Serano, Y.-M. Liu, and J. Weiner. Continuous, dense, highly collimated sodium beam. *Rev. Sci. Instr.*, 71: 3668–3676, 2000.

[107] Delves H. Effective range expansions of the scattering matrix. *Nucl. Phys.*, 8: 358–373, 1958.

[108] Desbiolles P., M. Arndt, P. Szriftgiser, and J. Dalibard. Elementary Sisyphus process close to a dielectric surface. *Phys. Rev. A*, 54: 4292–4298, 1996.

[109] Deutsch I. H. and G. K. Brennen. Quantum computing with neutral atoms in an optical lattice. *Fortschr. Phys.*, 48: 925–943, 2000.

[110] DiVincenzo. D. P. The physical implementation of quantum computation. *Fortschr. Phys.*, 48: 771–783, 2000.

[111] Dodd R. J., M. Edwards, C. J. Williams, C. W. Clark, M. J. Holland, P. A. Ruprecht, and K. Burnett. Role of attractive interactions on Bose–Einstein condensation. *Phys. Rev. A*, 54: 661–664, 1996.

[112] Doery M. R., E. J. D. Vredengbregt, J. G. C. Tempelaars, H. C. W. Beijerinck, and B. J. Verhaar. Long-range diatomic S+P potentials of heavy rare gases. *Phys. Rev. A*, 57: 3603–3620, 1998.

[113] Donley E. A., N. R. Claussen, S. L. Cornish, J. L. Roberts, E. A. Cornell, and C. E. Wieman. Dynamics of collapsing and exploding Bose–Einstein condensates. *Nature*, 412: 295–299, 2001.

[114] Donley E. A., N. R. Claussen, S. T. Thompson, and C. E. Wieman. Atom–molecule coherence in a Bose–Einstein condensate. *Nature*, 417: 529–533, 2002.

[115] Doyle J. M., B. Friedrich, J. Kim, and D. Patterson. Buffer-gas loading of atoms and molecules into a magnetic trap. *Phys. Rev. A*, 52: R2515–R2518, 1995.

[116] Drag C., B. L. Tolra, B. T'Jampens, D. Comparat, M. Allegrini, A. Crubellier, and P. Pillet. Photoassociative spectroscopy as a self-sufficient tool for the determination of the Cs triplet scattering length. *Phys. Rev. Lett.*, 85: 1408–1411, 2000.

[117] Drummond P. D., K. V. Kheruntsyan, D. J. Heinzen, and R. H. Wynar. Stimulated Raman adiabatic passage from an atomic to a molecular Bose–Einstein condensate.

[118] Dulieu O., P. D. Lett, K. Jones, U. Volz, C. Amiot, and F. Masnou-Seeuws. Interpretation of two-color photoassociation spectroscopy experiments in a cold sodium sample: evidence for a $^1\Pi_u$ doubly-excited autoinoizing state. Poster TU172 Abstracts of Contributed Papers, XX ICPEAC, Vienna, Vol. 2, 1997.

[119] Dulieu O., S. Magnier, and F. Masnou-Seeuws. Doubly-excited states for the $Na_2$ molecule: application to the dynamics of the associative ionization reaction. *Z. Phys. D*, 32: 229–240, 1994.

[120] Dulieu O., J. Weiner, and F. Masnou-Seeuws. On the accuracy of molecular data in the understanding of ultracold collisions. *Phys. Rev. A*, 49: 607–610, 1994.

[121] Dulieu O. A., A. Giusti-Suzor, and F. Masnou-Seeuws. Theoretical treatment of the associative ionization reaction between two laser-excited sodium atoms. Direct and indirect processes. *J. Phys. B*, 24: 4391–4408, 1991.

[122] Dulieu O. A. and P. S. Julienne. Coupled channel bound states calculations for alkali dimers using the Fourier grid method. *J. Chem Phys.*, 103: 60–66, 1995.

[123] Edwards M. and K. Burnett. Numerical solution of the nonlinear Schroedinger equation for small samples of trapped neutral atoms. *Phys. Rev. A*, 51: 1382–1386, 1995.

[124] Esry B., C. H. Greene, J. P. Burke, and J. L. Bohn. Hartree–Fock theory for double condensates. *Phys. Rev. Lett.*, 78: 3594–3597, 1997.

[125] Esry B., C. H. Greene, Y. Zhou, and C. D. Lin. Role of the scattering length in three-boson dynamics and Bose–Einstein condensation. *J. Phys. B*, 29: L51–L57, 1996.

[126] Esslinger T., I. Bloch, and T. W. Hänsch. Bose–Einstein condensation in a quadrupole-Ioffe-configuration trap. *Phys. Rev. A*, 58: R2664–R2667, 1998.

[127] Faulstich A., A. Schnetz, M. Sigel, T. Sleator, O. Carnal, V. Balykin, H. Takuma, and J. Mlynik. Strong velocity compression of a supersonic atomic beam using moving optical molasses. *Europhys. Lett.*, 17: 393–399, 1992.

[128] Fedichev P. O., Yu. Kagan, G. V. Shlyapnikov, and J. T. M. Walraven. Influence of nearly resonant light on the scattering length in low-temperature atomic gases. *Phys. Rev. Lett.*, 77: 2913–2916, 1996.

[129] Fedichev P. O., M. W. Reynolds, U. M. Rahmanov, and G. V. Shlyapnikov. Inelastic decay processes in a gas of spin-polarized triplet helium. *Phys. Rev. A*, 53: 1447–1453, 1996.

[130] Fedichev P. O., M. W. Reynolds, and G. V. Shlyapnikov. Three-body recombination of ultracold atoms to a weakly bound $s$ level. *Phys. Rev. Lett.*, 77: 2921–2924, 1996.

[131] Feng P., D. Hoffmann, and T. Walker. Comparison of trap-loss collision spectra for $^{85}$Rb and $^{87}$Rb. *Phys. Rev. A*, 47: R3495–R3498, 1993.

[132] Ferrari G., M. Inguscio, W. Jastrzebski, G. Modugno, G. Roati, and A. Simoni. Collisional properties of ultracold K–Rb mixtures. *Phys. Rev. Lett.*, 89: 053202-1–053202-4, 2002.

[133] Feynman R. P. *Feynman Lectures on Computation*. Perseus, 1996, edited by A. J. G Hey and R. Allen.

[134] Fioretti A., D. Comparat, A. Crubellier, O. Dulieu, F. Masnou-Seeuws, and P. Pillet. $Cs_2$ cold molecule formation through photoassociative scheme in a Cs vapor-cell magneto-optical trap. *Phys. Rev. Lett.*, 80: 4402–4405, 1998.

[135] Fioretti A., J. H. Müller, P. Verkerk, M. Allegrini, E. Arimondo, and P. S. Julienne. Direct measurement of fine-structure collisional losses from a Cs magneto optical trap. *Phys. Rev. A*, 55: R3999–R4002, 1997.

[136] Flemming J., A. M. Tuboy, D. M. B. P. Milori, L. G. Marcassa, S. C. Zilio, and V. S. Bagnato. Magneto-optical trap for sodium atoms operating on the D1 line. *Opt. Commun.*, 135: 269–272, 1997.

[137] Freeland R. S., C. C. Tsai, M. Marinescu, R. A. Cline, J. D. Miler, C. J. Williams, A. Dalgarno, and D. J. Heinzen. Rubidium atom lifetime and transition moment. University of Texas, Austin, Texas, 1997.

[138] Fried D. G., T. C. Killian, L. Willmann, D. Landhuis, S. C. Moss, D. Kleppner, and T. J. Greytak. Bose–Einstein condensation of atomic hydrogen. *Phys. Rev. Lett.*, 81: 3811–3814, 1998.

[139] Frisch C. R. Experimenteller Nachweis des Einsteinschen Strahlungsrückstosses. *Z. Phys.*, 86: 42–48, 1933.

[140] Gabbanini C., A. Fioretti, A. Lucchesini, S. Gozzini, and M. Mazzoni. Cold rubidium molecules formed in a magneto-optical trap. *Phys. Rev. Lett.*, 84: 2814–2817, 2000.

[141] Gadéa F. X., T. Leininger, and A. S. Dickinson. Accurate calculation of the scattering length for the cooling of hydrogen atoms by lithium atoms. *Eur. Phys. J. D*, 15: 251–255, 2001.

[142] Gadéa F. X., T. Leininger, and A. S. Dickinson. Calculated scattering length for spin-polarized metastable helium. *J. Chem. Phys.*, 117: 7122–7127, 2002.

[143] Gallagher A. Associative ionization in collisions of slowed and trapped sodium. *Phys. Rev. A*, 44: 4249–4259, 1991.

[144] Gallagher A. and D. E. Pritchard. Exoergic collisions of cold Na*–Na. *Phys. Rev. Lett.*, 63: 957–960, 1989.

[145] Gao B. Theory of slow-atom collisions. *Phys. Rev. A*, 54: 2022–2039, 1997.

[146] Gardner J. R., R. A. Cline, J. D. Miller, D. J. Heinzen, H. M. J. M. Boesten, and B. J. Verhaar. Collisions of doubly spin-polarized, ultracold $^{85}$Rb atoms. *Phys. Rev. Lett.*, 74: 3764–3767, 1995.

[147] Gaupp A., P. Kuske, and H. J. Andrä. Accurate lifetime measurements of the lowest $^2P_{1/2}$ states in neutral lithium and sodium. *Phys. Rev. A*, 26: 3351–3359, 1982.

[148] Geltman S. and A. Bambini. Triplet scattering lengths for rubidium and their role in Bose–Einstein condensation. *Phys. Rev. Lett.*, 86: 3276–3279, 2001.

[149] Gensemer S. D. and P. L. Gould. Ultracold collisions observed in real time. *Phys. Rev. Lett.*, 80: 936–939, 1997.

[150] Gensemer S. D., V. Sanchez-Villicana, K. Y. N. Tan, T. T. Grove, and P. L. Gould. Trap-loss collisions of $^{85}$Rb and $^{87}$Rb: dependence on trap parameters. *Phys. Rev. A*, 56: 4055–4063, 1997.

[151] Ghezali S., Ph. Laurent, S. N. Lea, and A. Clairon. An experimental study of the spin-exchange frequency shift in a laser-cooled cesium fountain frequency standard. *Europhys. Lett.*, 36: 25–30, 1996.

[152] Gibble K., S. Chang, and R. Legere. Direct observation of s-wave atomic collisions. *Phys. Rev. Lett.*, 75: 2666–2669, 1995.

[153] Gibble K. and S. Chu. Future slow-atom frequncy standards. *Metrologia*, 29: 201–212, 1992.

[154] Gibble K. and S. Chu. Laser-cooled Cs frequency standard and a measurement of the frequency shift due to ultracold collisions. *Phys. Rev. Lett.*, 70: 1771–1774, 1993.

[155] Gibble K. and B. J. Verhaar. Eliminating cold-collision freuency shifts. *Phys. Rev. A*, 52: 3370–3373, 1995.

[156] Gordon J. P. and A. Ashkin. Motion of atoms in a radiation trap. *Phys. Rev. A*, 21: 1606–1617, 1980.

[157] Gott H. V., M. S. Ioffe, and V. G. Telkovsky. *Nuclear Fusion*. Supplement Part 3. International Atomic Energy Agency, Vienna, 1962. p. 1405.

[158] Gould P. L., P. D. Lett, P. S. Julienne, W. D. Phillips, H. R. Thorsheim, and J. Weiner. Observation of associative ionization of ultracold laser-trapped sodium atoms. *Phys. Rev. Lett.*, 60: 788–791, 1988.

[159] Gribakin G. F. and V. V. Flambaum. Calculation of the scattering length in atomic collisions using the semiclassical approximation. *Phys. Rev. A*, 48: 546–553, 1993.

[160] Guéry-Odelin D., J. Söding, P. Desbiolles, and J. Dalibard. Is Bose–Einstein condensation of atomic cesium possible? *Europhys. Lett.*, 44: 25–30, 1998.

[161] Hadzibabic Z., C. A. Stan, K. Dieckmann, S. Gupta, M. W. Zwierlein, A. Görlitz, and W. Ketterle. Two-species mixture of quantum degenerate Bose and Fermi gases. *Phys. Rev. Lett.*, 88: 160401-1–16401-4, 2002.

[162] Hammes M., D. Rychtarik, B. Engeser, H.-C. Nägerl and R. Grimm. Evanescent-wave trapping and evaporative cooling of an atomic gas near two-dimensionality. *LANL e-print arXiv:physics\0208065*, v1: 1–4, August 2002.

[163] Hammes M., D. Rychtarik, H.-C. Nägerl, and R. Grimm. Cold atom gas at very high densities in an optical surface microtrap. *LANL e-print arXiv:physics\0204026*, v1: 1–4, April 2002.

[164] Han D.-J., A. Rynar, Ph. Courteille, and D. J. Heinzen. Bose–Einstein condensation of large numbers of $^{87}$Rb atoms in a magnetic TOP trap. *Phys. Rev. A*, 57: 958–975, 1997.

[165] Harris S. E. Control of Feshbach resonances by quantum interference. *Phys. Rev. A*, 66: 010701-1–010701-4, 2002.

[166] Heather R. W. and P. S. Julienne. Theory of laser-induced associative ionization of ultracold Na. *Phys. Rev. A*, 47: 1887–1906, 1993.

[167] Heinzen D. J. *Atomic Physics*, volume 14, pages 369–388. American Institute of Physics, New York, 1995. Collisions of Ultracold Atoms in Optical Fields.

[168] Heinzen D. J., R. Wynar, P. D. Drummond, and K. V. Kheruntsyan. Superchemistry: dynamics of coupled atomic and molecular Bose–Einstein condensates. *Phys. Rev. Lett.*, 84: 5029–5033, 2000.

[169] Henriet A., F. Masnou-Seeuws, and O. Dulieu. Diabatic representation of the excited states of the Na$_2$ molecule: application to the associative ionization reaction between two excited sodium atoms. *Z. Phys. D*, 18: 287–298, 1991.

[170] Herschbach D. R. Molecular beams. In J. Ross, editor, *Advances in Chemical Physics*, volume X, pages 319–393. Wiley Interscience, New York, 1966.

[171] Herzberg G. *Molecular Spectra and Molecular Structure I, Spectra of Diatomic Molecules*, volume I. van Nostrand, Princeton, 2nd edition, 1950.

[172] Hess H. F. Evaporative cooling of magnetically trapped and compressed spin-polarized hydrogen. *Phys. Rev. B*, 34: 3476–3479, 1986.

[173] Hess H. F., G. P. Kochanski, J. M. Doyle, T. J. Greytak, and D. Kleppner. Spin-polarized hydrogen maser. *Phys. Rev. A*, 34: 1602–1604, 1986.

[174] Hess H. F., G. P. Kochanski, J. M. Doyle, N. Mashuhara, D. Kleppner, and T. J. Greytak. Magnetic trapping of spin-polarized atomic hydrogen. *Phys. Rev. Lett.*, 59: 672–675, 1987.

[175] Hoffmann D., S. Bali, and T. Walker. Trap-depth measurements using ultracold collisions. *Phys. Rev. A*, 54: R1030–R1033, 1996.

[176] Hoffmann D., P. Feng, and T. Walker. Measurements of Rb trap-loss collision spectra. *J. Opt. Soc. Am. B*, 11: 712, 1994.

[177] Hoffmann D., P. Feng, R. S. Williamson, and T. Walker. Excited-state collision of trapped $^{85}$Rb atoms. *Phys. Rev. Lett.*, 69: 753–756, 1992.

[178] Holland M. J., D. S. Jin, M. L. Chiofalo, and J. Cooper. Emergence of interaction effects in Bose–Einstein condensation. *Phys. Rev. Lett.*, 78: 3801–3805, 1997.

[179] Holland M. J., K.-A. Suominen, and K. Burnett. Cold collisions in a laser field: quantum Monte Carlo treatment of radiative heating. *Phys. Rev. A*, 50: 1513–1530, 1994.

[180] Hoogerland. M. PhD thesis, Eindhoven University, 1993.

[181] Hopkins S. A., W. Webster, J. Arlt, P. Bance, S. Cornish, O. Marago, and C. J. Foot. Measurement of elastic cross section for cold cesium collisions. *Phys. Rev. A*, 61: 032707-1–032707-4, 2000.

[182] Huang K. and C. N. Yang. Quantum mechanical many-body problem with hard-sphere interactions. *Phys. Rev.*, 105: 767–775, 1957.

[183] Huennekens J. and A. Gallagher. Associative ionization in collisions between two Na(3P) atoms. *Phys. Rev. A*, 28: 1276–1287, 1983.

[184] Hurlimann M. D., W. N. Hardy, A. J. Berlinski, and R. W. Cline. Recirculating cryogenic hydrogen maser. *Phys. Rev. A*, 34: 1605–1608, 1986.

[185] Huynh B., O. Dulieu, and F. Masnou-Seeuws. Associative ionization between two laser-excited sodium atoms: theory compared to experiment. *Phys. Rev. A*, 57: 958–975, 1998.

[186] Inoue G., J. K. Ku, and D. W. Setser. Photoassociative laser induced fluorescence of XeCl. *J. Chem. Phys.*, 76: 733–734, 1982.

[187] Inouye S., M. R. Andrews, J. Stneger, H.-J. Miesner, D. M. Stamper-Kurn, and W. Ketterle. Observation of Feshbach resonances in a Bose–Einstein condensate. *Nature*, 392: 151–154, 1998.

[188] Jaksch D., H.-J. Briegel, J. I. Cirac, C. W. Gardiner, and P. Zoller. Entaglement of atoms via cold controlled collisions. *Phys. Rev. Lett.*, 82: 1975–1978, 1999.

[189] Jamieson M. J., A. Dalgarno, and J. M. Doyle. Scattering lengths for collisions of ground state and metastable state hydrogen atoms. *Mol. Phys.*, 87: 817–826, 1996.

[190] Javanainen J. and M. Mackie. Coherent photoassociastion of a Bose–Einstein condensate. *Phys. Rev. A*, 59: R3186–R3189, 1999.

[191] Javanainen J. and M. Mackie. Rate limit for photoassociation of a Bose–Einstein condensate. *Phys. Rev. Lett.*, 88: 090403-1–090403-4, 2002.

[192] Jessen P. S. and I. H. Deutsch. Optical lattices. *Adv. At. Mol. Opt. Phys.*, 37: 95–138, 1996.

[193] Joachain C. J. *Quantum Collision Theory*. North-Holland, The Netherlands, 1975.

[194] Jones K. M., P. S. Julienne, P. D. Lett, W. D. Phillips, E. Tiesinga, and C. J. Williams. Measurement of the atomic Na(3P) lifetime and of retardation in the interaction between two atoms bound in a molecule. *Europhys. Lett.*, 35: 85–90, 1996.

[195] Jones K. M., S. Maleki, S. Bize, P. D. Lett, C. J. Williams, H. Richling, H. Knöckel, E. Tiemann, H. Wang, P. L. Gould, and W. C. Stwalley. Direct measurement of the ground-state dissociation energy of $Na_2$. *Phys. Rev. A*, 54: R1006–R1009, 1997.

[196] Jones K. M., S. Maleki, L. P. Ratliff, and P. D. Lett. Two-colour photoassociation spectroscopy of ultracold sodium. *J. Phys. B*, 30: 289–308, 1997.

[197] Jones R. B., J. H. Schloss, and J. G. Eden. Excitation spectra for the photoassociation of Kr–F and Xe–I collision pairs in the ultraviolet (209–258) nm. *J. Chem. Phys.*, 98: 4317–4334, 1993.

[198] Julienne P. S. Laser modification of ultracold atomic collisions in optical traps. *Phys. Rev. Lett.*, 61: 698–701, 1988.

[199] Julienne P. S. Cold binary atomic collisions in a light field. *J. Res. Natl. Inst. Stand. Technol.*, 101: 487–503, 1996.

[200] Julienne P. S., K. Burnett, Y. B. Band, and W. C. Stwalley. stimulated Raman molecule production in Bose–Einstein condensates. *Phys. Rev. A*, 58: R797–R800, 1998.

[201] Julienne P. S. and R. Heather. Laser modification of ultracold atomic collisions: theory. *Phys. Rev. Lett.*, 67: 2135–2138, 1991.

[202] Julienne P. S. and F. H. Mies. Collisions of ultracold trapped atoms. *J. Opt. Soc. Am. B*, 6: 2257–2269, 1989.

[203] Julienne P. S., F. H. Mies, E. Tiesinga, and C. J. Williams. Collisional stability of double Bose condensates. *Phys. Rev. Lett.*, 78: 1880–1883, 1997.

[204] Julienne P. S., A. M. Smith, and K. Burnett. Theory of collisions between laser cold atoms. *Adv. At. Mol. Opt. Phys.*, 30: 141–198, 1993.

[205] Julienne P. S., K.-A. Suominen, and Y. Band. Complex-potential model of collisions of laser-cooled atoms. *Phys. Rev. A*, 49: 3890–3896, 1995.

[206] Julienne P. S. and J. Vigué. Cold collisions of ground- and excited-state alkali-metal atoms. *Phys. Rev. A*, 44: 4464–4485, 1991.

[207] Julienne P. S., C. Williams, O. Dulieu, and Y. B. Band. Calculations of collisional loss rates of trapped Li atoms. *Laser Phys.*, 4: 1076–1084, 1994.

[208] Kagan Yu., A. E. Muryshev, and G. V. Shlyapnikov. Collapse and Bose–Einstein condensation in a trapped Bose gas with negative scattering length. *Phys. Rev. Lett.*, 81: 933–937, 1998.

[209] Kagan Yu., B. V. Svistunov, and G. V. Shlyapnikov. Effect of Bose–Einstein condensation on inelastic processes in gases. *Pis. Zh. Eksp Teor. Fiz.*, 42: 169–172, 1985. (JETP Lett. **42**, 209–212).

[210] Kasevich M. A., E. Riis, S. Chu, and R. G. DeVoe. Rf spectroscopy in an atomic fountain. *Phys. Rev. Lett.*, 63: 612–615, 1989.

[211] Katori H., Kunugita, and T. Ido. Quantum statistical effect on ionizing collisions of ultracold metastable Kr isotopes. *Phys. Rev. A*, 52: R4324–R4327, 1995.

[212] Katori H. and F. Shimizu. Laser-induced ionizing collisions of ultracold Krypton gas in the $1s_5$ metastable state. *Phys. Rev. Lett.*, 73: 2555–2558, 1994.

[213] Kawanaka J., K. Shimizu, H. Takuma, and F. Shimizu. Quadratic collisional loss rate of a $^7$Li trap. *Phys. Rev. A*, 48: R883–R885, 1993.

[214] Kazanstsev A. P., G. I. Surdutovich, D. O. Chudesnikov, and V. P. Yakovlev. Scattering, velocity bunching and self-localization of atoms in a light field. *J. Opt. Soc. Am. B*, 6:2130–2139, 1989.

[215] Ketterle W., K. B. Davis, M. A. Joffe, A. Martin, and D. Pritchard. High densities of cold atoms in a dark spontaneous-force optical trap. *Phys. Rev. Lett.*, 70: 2253–2256, 1993.

[216] Ketterle W. and N. J. Van Druten. Evaporative cooling of trapped atoms. *Adv. At. Mol. Opt. Phys.*, 37: 181–236, 1996.

[217] Ketterle W. and H.-J. Miesner. Coherence properties of Bose condensates and atom lasers. *Phys. Rev. A*, 56: 3291–3293, 1997.

[218] Kim K. H., K. I. Lee, H. R. Noh, W. Jhe, N. Kwon, and M. Ohtsu. Cold atomic beam produced by a conical mirror funnel. *Phys. Rev. A*, 64: 013402-1–013402-5, 2001.

[219] King G. W. and J. H. Van Vleck. Dipole–dipole resonance forces. *Phys. Rev.*, 55: 1165–1172, 1939.

[220] Kleppner D., H. C. Berg, S. B. Crapmton, N. F. Ramsey, R. C. Vessot, H. E. Peters, and J. Vanier. Hydrogen-maser principles and techniques. *Phys. Rev.*, 138: A972–A983, 1965.

[221] Koelman J. M. V. A., S. B. Crampton, H. T. C. Stoof, O. J. Luiten, and B. J. Verhaar. Spin polarized deuterium in magnetic traps. *Phys. Rev. A*, 38: 3535–3547, 1988.

[222] Koelman J. M. V. A., H. T. C. Stoof, B. J. Verhaar, and J. T. M. Walraven. Spin polarized deuterium in magnetic traps. *Phys. Rev. Lett.*, 59: 676–679, 1987.

[223] Koelman J. M. V. A., H. T. C. Stoof, B. J. Verhaar, and J. T. M. Walraven. Lifetime of magnetically trapped ultracold atomic deuterium gas. *Phys. Rev. B*, 38: 9319–9322, 1988.

[224] Köhler T. and K. Burnett. Microscopic quantum dynamics approach to the dilute condensed bose gas. *Phys. Rev. A*, 65: 033601-1–033601-8, 2001.

[225] Köhler T., T. Gaesenzer, and K. Burnett. Microscopic theory of atom–molecule oscillations in a Bose–Einstein condensate. *LANL e-print arXiv cond-mat\0209100*, v1: 1–18, 4 Sept 2002.

[226] Kokkelmans S. J. J. M. F., H. M. J. M. Boesten, and B. J. Verhaar. Role of collisions in creation of overlapping Bose condensates. *Phys. Rev. A*, 55: R1589–R1592, 1997.

[227] Kokkelmans S. J. J. M. F., B. J. Verhaar, and K. Gibble. Prospects for Bose–Einstein condensation in cesium. *Phys. Rev. Lett.*, 81: 951–954, 1998.

[228] Kokkelmans S. J. J. M. F., B. J. Verhaar, K. Gibble, and D. J. Heinzen. Predictions for laser-cooled Rb clocks. *Phys. Rev. A*, 56: R4389–R4392, 1997.

[229] Kokkelmans S. J. J. M. F. unpublished private communication, the results of which are plotted in Fig. 5 of [114], 2002.

[230] Koštrun M., M. Mackie, R. Côté, and J. Javanainen. Theory of coherent photoassociation of Bose–Einstein condensate. *Phys. Rev. A*, 62: 063616-1–063616-23, 2000.

[231] Krauss M. and W. J. Stevens. Effective core potentials and accurate energy curves for $Cs_2$ and other alkali diatomics. *Chem. Phys.*, 93: 4236–4242, 1990.

[232] Kunugita H., T. Ido, and F. Shimizu. Ionizing collision rate of metastable rare-gas atoms in an optical lattice. *Phys. Rev. Lett.*, 79: 621–624, 1997.

[233] Kuyatt C. E. Electron optics lectures. available from the author at NIST, 1967.

[234] Lagendijk A., I. F. Silvera, and B. J. Verhaar. Spin-exchange and dipolar relazation rates in atomic hydrogen: lifetimes in magnetic traps. *Phys. Rev. B*, 33: 626–628, 1986.

[235] Lai W. K., K.-A. Suominen, B. M. Garraway, and S. Stenholm. Dissipation effects on wave packets in level crossings: a comparison between two numerical approaches. *Phys. Rev. A*, 47: 4779–4785, 1993.

[236] Laloë. F. Spin polarized quantum systems. *J. Phys. (Paris) Suppl.*, Colloque C7, 1980. F. Laloë, editor.

[237] Lawall J., C. Orzel, and S. L. Rolston. Suppression and enhancement of collisions in optical lattices. *Phys. Rev. Lett.*, 80: 480–483, 1998.

[238] Lee T. D. and C. N. Yang. Low-temperature behavior of dilute Bose system of hard spheres I. Equilibrium properties. *Phys. Rev.*, 112: 1419–1429, 1958.

[239] Leo P., E. Tiesinga, P. S. Julienne, D. K. Walter, S. Kadlecek, and T. G. Walter. Elastic and inelastic collisions of cold spin-polarized [113]Cs atoms. *Phys. Rev. Lett.*, 81: 1389–1392, 1998.

[240] Leo P. J., C. J. Williams, and P. S. Julienne. Collision properties of ultracold [133]Cs atoms. *Phys. Rev. Lett.*, 85: 2721–2724, 2000.

[241] Leonhardt D. and J. Weiner. Direct two-color photoassociative ionization in a rubidium magneto-optic trap. *Phys. Rev. A*, 52: 1419–1429, 1995. See also erratum, *Phys. Rev. A*, 53: 2904(E), 1996.

[242] LeRoy R. J. and R. B. Bernstein. Dissociation energy and long-range ptoential of diatomic molecules from vibrational spacings of higher levels. *J. Chem. Phys.*, 52: 3869–3879, 1970.

[243] Lett P. D., K. Helmerson, W. D. Phillips, L. P. Ratliff, S. L. Rolston, and M. E. Wagshul. Spectrsocopy of $Na_2$ by photoassociation of laser-cooled Na. *Phys. Rev. Lett.*, 71: 2200–2203, 1993.

[244] Lett P. D., P. S. Jessen, W. D. Phillips, S. L. Rolston, C. I. Westbrook, and P. L. Gould. Laser modification of ultracold collisions: experiment. *Phys. Rev. Lett.*, 67: 2139–2142, 1991.

[245] Lett P. D., P. S. Julienne, and W. D. Phillips. Photoassociative spectroscopy of laser cooled atoms. *Annu. Rev. Phys. Chem.*, 46: 423–452, 1996.

[246] Lett P. D., K. Mølmer, S. D. Gensemer, K. Y. N. Tan, A. Kumarakrishnan, C. D. Wallace, and P. L. Gould. Hyperfine structure modifications of collisional losses from light-force atom traps. *J. Phys. B*, 28: 65–81, 1995.

[247] Lett P. D., R. N. Watts, C. I. Westbrook, W. D. Phillips, P. L. Gould, and H. J. Metcalf. Observation of atoms, laser-cooled below the Doppler limit. *Phys. Rev. Lett.*, 61: 169–172, 1988.

[248] Lovelace R. V. E., C. Mehanian, J. J. Tommila, and D. M. Lee. Magnetic confinement of a neutral gas. *Nature*, 318: 30–36, 1985.

[249] Lu Z. T., K. L. Corwin, M. J. Renn, M. H. Anderson, E. A. Cornell, and C. E. Wieman. Low-velocity intense source of atoms from a magneto-optical trap. *Phys. Rev. Lett.*, 77: 3331–3334, 1996.

[250] Maan A. C., H. T. C. Stoof, and B. J. Verhaar. Cryogenic *H* maser in a strong *B* field. *Phys. Rev. A*, 41: 2614–2620, 1990.

[251] Mackie M. Feshbach-stimulated photoproduction of a stable molecular condensate. *arXiv e-print*, 3(physics/0202041): 1–5, June 2002.

[252] Mackie M., K.-A. Suominen, and J. Javanainen. Mean-field theory of Feshbach-resonant interactions in $^{85}$Rb condensates. *LANL e-print arXiv cond-mat\0205535, v2*: 1–5, 30 July 2002.

[253] Maddi J. A., T. P. Dinneen, and H. Gould. Slowing and cooling molecules and neutral atoms by time-varying electric-field gradients. *Phys. Rev. A*, 60: 3882–3891, 1999.

[254] Magnier S., Ph. Millié, O. Dulieu, and F. Manous-Seeuws. Potential curves for the ground and excited states for the Na$_2$ molecule up to the $(3s + 5p)$ dissociation limit: results of two different effective potential calculations. *J. Chem. Phys.*, 98: 7113–7125, 1993.

[255] Mancini M. W., A. L. de Oliveira, K. M. F. Magalhães, V. S. Bagnato, and L. G. Marcassa. Intensity and detuning dependence of fine structure changing collisions in a $^{85}$Rb magneto-optical trap. *Eur. Phys. J. D*, 13: 317–322, 2001.

[256] Maracassa L. G., G. D. Telles, and S. R. Muniz. Photoassociative ionization using two independent colors in sodium-vapor-cell magneto-optical traps. *Phys. Rev. A*, 60: 1305–1310, 1999.

[257] Marcassa L., V. Bagnato, Y. Wang, C. Tsao, J. Weiner, O. Dulieu, Y. B. Band, and P. S. Julienne. Collisional loss rate in a magneto-optical trap for sodium atoms: light intensity dependence. *Phys. Rev. A*, 47: R4563–R4566, 1993.

[258] Marcassa L., K. Helmerson, A. M. Tuboy, D. M. B. P. Milori, S. R. Muniz, J. Flemming, S. C. Zilio, and V. S. Bagnato. Collisional loss rate of sodium atoms in a magneto-optical trap operating on the D1 line. *J. Phys. B*, 29: 3051–3057, 1996.

[259] Marcassa L., R. Horowicz, X. Zilio, V. Bagnato, and J. Weiner. Intensity dependence of optical suppression in photoassociative ionization collisions in a sodium magneto-optical trap. *Phys. Rev. A*, 52: R913–R916, 1995.

[260] Marcassa L., S. Muniz, E. de Queiroz, S. Zilio, V. Bagnato, J. Weiner, P. S. Julienne, and K.-A. Suominen. Optical suppression of photoassociative ionization in a magneto-optical trap. *Phys. Rev. Lett.*, 73: 1911–1914, 1994.

[261] Marcassa L., S. R. Muniz, and J. Flemming. Optical catalysis in a Na-vapor cell MOT. *Braz. J. Phys.*, 27: 238–242, 1997.

[262] Marcassa L. G., G. D. Telles, S. R. Muniz, and V. S. Bagnato. Collisional losses in a K–Rb cold mixture. *Phys. Rev. A*, 63: 013413-1–013413-6, 2000.

[263] Marcassa L. G., R. A. S. Zanon, S. Dutta, J. Weiner, O. Dulieu, and V. S. Bagnato. Direct measurement of fine structure changing collisional losses in cold trapped $^{85}$Rb. *Eur. Phys. J. D*, 7: 317–321, 1999.

[264] Mastwijk H. C., J. W. Thomsen, P. van der Straten, and A. Niehaus. Optical collisions of cold, metastable helium atoms. *Phys. Rev. Lett.*, 80: 5516–5519, 1998.

[265] Masuhara N., J. M. Doyle, J. C. Sandberg, D. Kleppner, and T. J. Greytak. Evaporative cooling of spin-polarized atomic hydrogen. *Phys. Rev. Lett.*, 61: 935–938, 1988.

[266] McAlexander W. I., E. R. I. Abraham, N. W. M. Ritchie, C. J. Williams, H. T C. Stoof, and R. G. Hulet. Precise atomic radiative lifetime via phtoassociative spectroscopy of ultracold lithium. *Phys. Rev. A*, 51: R871–R874, 1995.

[267] McAlexander W. I., E. R. I. Abraham, N. W. M. Ritchie, C. J. Williams, H. T. C. Stoof, and R. G. Hulet. Radiative lifetime of the 2P state of lithium. *Phys. Rev. A*, 54: R5–R8, 1996.

[268] McKenzie C., J. H. Denschlag, H. Häffner, A. Browaeys, L. E. E. de Araujo, F. K. Fatemi, K. M. Jones, J. E. Simsarian, D. Cho, A. Simoni, E. Tiesinga, P. S. Julienne, K. Helmerson, P. D. Lett, S. L. Rolston, and W. D. Phillips. Photoassociation of sodium in a Bose–Einstein condensate. *Phys. Rev. Lett.*, 88: 120403-1–120403-4, 2002.

[269] McLone R. R. and E. A. Power. On the interaction between two identical neutral dipole systems, one in an excited state and the other in the ground state. *Mathematika*, 11: 91–94, 1964.

[270] Meath W. J. Retarded interaction energies between like atoms in different energy states. *J. Chem. Phys.*, 48: 227–235, 1968.

[271] Metcalf H. and P. van der Straten. Cooling and trapping of neutral atoms. *Phys. Rep.*, 224: 203–286, 1994.

[272] Mewes M.-O., M. R. Andrews, D. M. Kurn, D. S. Durfee, C. G. Townsend, and W. Ketterle. Output coupler for Bose–Einstein condensed atoms. *Phys. Rev. Lett.*, 78: 582–585, 1997.

[273] Mewes M.-O., M. R. Andrews, N. J. van Druten, D. M. Kurn, D. S. Durfee, and W. Ketterle. Bose–Einstein condensation in a tightly confining dc magnetic trap. *Phys. Rev. Lett.*, 77: 416–419, 1996.

[274] Mies F. H. Molecular theory of atomic collisions: calculated cross section for $H^+ + F(^2 P)$. *Phys. Rev. A*, 7: 957–967, 1973.

[275] Mies F. H. Molecular theory of atomic collisions: fine-structure transitions. *Phys. Rev. A*, 7: 942–957, 1973.

[276] Mies F. H., E. Tiesinga, and P. S. Julienne. Manipulation of Feshbach resonances in ultracold aotmic collisions using time-dependent magnetic fields. *Phys. Rev. A*, 61: 022721-1–022721-17, 2000.

[277] Mies F. H., C. J. Williams, P. S. Julienne, and M. Krauss. Estimating bounds on collisional relaxation rates of spin-polarized $^{87}$Rb atoms at ultracold temperatures. *J. Res. Natl. Inst. Stand. Technol.*, 101: 521–535, 1996.

[278] Migdall A. L., J. V. Prodan, W. D. Phillips, T. H. Bergman, and H. J. Metcalf. First observation of magnetically trapped neutral atoms. *Phys. Rev. Lett.*, 54: 2596–2599, 1985.

[279] Miller J. D., R. A. Cline, and D. J. Heinzen. Far-off-resonance optical trapping of atoms. *Phys. Rev. A*, 47: R4567–R4570, 1993.

[280] Miller J. D., R. A. Cline, and D. J. Heinzen. Photoassociation spectrum of ultracold Rb atoms. *Phys. Rev. Lett.*, 71: 2204–2207, 1993.

[281] Modugno G., G. Ferrari, G. Roati, R J. Brecha, A. Simoni, and M. Inguscio. Bose–Einstein condensation of potassium atoms by sympathetic cooling. *Science*, 294: 1320–1322, 2001.

[282] Moerdijk A. J., H. M. J. M. Boesten, and B. J. Verhaar. Decay of trapped ultracold alkali atoms by recombination. *Phys. Rev. A*, 53: 916–920, 1996.

[283] Moerdijk A. J., W. C. Stwalley, R. G Hulet, and B. J. Verhaar. Negative scattering length of ultracold $^7$Li gas. *Phys. Rev. Lett.*, 72: 40–43, 1994.

[284] Moerdijk A. J. and B. J. Verhaar. Prospects for Bose–Einstein condensation in atomic $^7$Li and $^{23}$Na. *Phys. Rev. Lett.*, 73: 518–521, 1994.

[285] Moerdijk A. J. and B. J. Verhaar. Laser cooling and the highest bound states of the Na diatom system. *Phys. Rev. A*, 51: R4333–R4336, 1995.

[286] Moerdijk A. J. and B. J. Verhaar. Collisional two- and three-body decay rates of dilute quantum gases at ultralow temperatures. *Phys. Rev. A*, 53: R19–R22, 1996.

[287] Moerdijk A. J., B. J. Verhaar, and A. Axelsson. Resonance in ultracold collisions of $^6$Li, $^7$Li, and $^{23}$Na. *Phys. Rev. A*, 51: 4852–4861, 1995.

[288] Moerdijk A. J., B. J. Verhaar, and T. M. Nagtegaal. Collisions of dressed ground-state atoms. *Phys. Rev. A*, 53: 4343–4351, 1996.

[289] Moi L. Application of a very long cavity laser to atom slowing down and optical pumping. *Opt. Commun.*, 50: 349–352, 1984.

[290] Molenaar P. A., P. van der Straten, and H. G. M. Heidemann. Long-range predissociation in two-color photoassociation of ultracold Na atoms. *Phys. Rev. Lett.*, 77: 1460–1463, 1996.

[291] Monroe. C. PhD thesis, University of Colorado, 1992.

[292] Monroe C., W. Swann, H. Robinson, and C. Wieman. Very cold trapped atoms in a vapor cell. *Phys. Rev. Lett.*, 65: 1571–1574, 1990.

[293] Monroe C. R., E. A. Cornell, C. A. Sackett, C. J. Myatt, and C. E. Wieman. Measurement of Cs–Cs elastic scattering at $T = 30\,\mu$K. *Phys. Rev. Lett.*, 70: 414–417, 1993.

[294] Morinaga M., M. Yasuda, T. Kishimoto, and F. Shimizu. Holographic manipulation of a cold atomic beam. *Phys. Rev. Lett.*, 77: 802–805, 1996.

[295] Mott N. F. and H. S. W. Massey. *The Theory of Atomic Collisions*. Oxford, Clarendon, 3rd edition, 1965.

[296] Mourachko I., D. Comparat, F. de Tomasi, A. Fioretti, P. Nosbaum, V. M. Akulin, and P. Pillet. Many-body effects in a frozen Rydberg gas. *Phys. Rev. Lett.*, 80: 253–256, 1998.

[297] Movre M. and G. Pichler. Resonance interaction and self-broadening of alkali resonance lines I. Adiabatic potentials curves. *J. Phys. B*, 10: 2631–2638, 1977.

[298] Muniz S. R., L. G. Marcassa, R. Napolitano, G. D. Telles, J. Weiner, S. C. Zilio, and V. S. Bagnato. Optical suppression of hyperfine-changing collisions in a sample of ultracold sodium atoms. *Phys. Rev. A*, 55: 4407–4411, 1997.

[299] Myatt C. J., E. A. Burt, R. W. Ghrist, E. A. Cornell, and C. E. Wieman. Production of two overlapping Bose–Einstein condensates by sympathetic cooling. *Phys. Rev. Lett.*, 78: 586–589, 1997.

[300] Myatt C. J., N. R. Newbury, R. W. Ghrist, S. Loutzenhiser, and C. E. Wieman. Multiply loaded magneto-optical trap. *Opt. Lett.*, 21: 290–292, 1996.

[301] Napolitano R., J. Weiner, and P. S. Julienne. Theory of optical suppression of ultracold-collision rates by polarized light. *Phys. Rev. A*, 55: 1191–1207, 1997.

[302] Napolitano R., J. Weiner, C. J. Williams, and P. S. Julienne. Line shapes of high resolution photoassociation spectra of optically cooled atoms. *Phys. Rev. Lett.*, 73: 1352–1355, 1994.

[303] Nellessen J., J. Werner, and W. Ertmer. Magneto-optical compression of a monoergetic sodium atomic beam. *Opt. Commun.*, 78: 300–308, 1990.

[304] Nesnidal R. C. and T. G. Walker. Light-induced ultracold spin-exchange collisions. *Phys. Rev. A*, 62: 030701-1–030701-4, 2000.

[305] Newbury N. R., C. J. Myatt, and C. E. Wieman. s-wave elastic collisions between cold ground state $^{87}$Rb atoms. *Phys. Rev. A*, 51: R2680–R2683, 1995.

[306] Nikolov A. N., E. E. Eyler, H. Wang, W. C. Stwalley, and P. L. Gould. Efficient production of ground-state potassium molecules at sub-mK tempertures by two-step photoassociation. *Phys. Rev. Lett.*, 84: 246–249, 2000.

[307] Orzel C., S. D. Bergeson, S. Kulin, and S. L. Rolston. Time-resolved studies of ultra-cold ionizing collisions. *Phys. Rev. Lett.*, 80: 5093–5096, 1998.

[308] OSA. *Photoassociative Spectroscopy of a Laser Cooled Binary Mixture*, OSA Technical Digest Series **12**, 85, 1997.

[309] Ovchinnikov Yu. B., I. Manek, and R. Grimm. Surface trap for Cs atoms based on evanescent-wave cooling. *Phys. Rev. Lett.*, 79: 2225–2228, 1997.

[310] Pereira Dos Santos F., J. Wang, C. J. Barrelet, F. Perales, E. Rasel, C. S. Unnikrishnan, M. Leduc, and C. Cohen-Tannoudji. Bose–Einstein condensation of metastable helium. *Phys. Rev. Lett.*, 86: 3459–3462, 2001.

[311] Peters M. G., D. Hoffmann, J. D. Tobiason, and T. Walker. Laser-induced ultracold $Rb(5S_{1/2})+Rb(5P_{1/2})$ collisions. *Phys. Rev. A*, 50: R906–R909, 1994.

[312] Phillips W. D., J. V. Prodan, and H. J. Metcalf. Laser cooling and electromagnetic trapping of neutral atoms. *J. Opt. Soc. Am. B*, 2: 1751–1767, 1985.

[313] Pierce J. R. *Theory and Design of Electron Beams*. Van Nostrand, Princton, New Jersey, 2nd edition, 1954.

[314] Pillet P., A. Crubeillier, A. Bleton, O. Dulieu, P. Nosbaum, I. Mourachko, and F. Masnou-Seeuws. Photoassociation in a gas of cold alklai atoms: I. Perturbative quantum approach. *J. Phys. B*, 30: 2801–2820, 1997.

[315] Prentiss M., A. Cable, J. Bjorkholm, S. Chu, E. Raab, and D. Pritchard. Atomic-density-dependent losses in an optical trap. *Opt. Lett.*, 13: 452–454, 1988.

[316] Pritchard D. E. Cooling netural atoms in a magnetic trap for precision spectroscopy. *Phys. Rev. Lett.*, 51: 1336–1339, 1983.

[317] Pritchard D. E. *Electron and Atomic Collisions*, pp. 593–604. North-Holland, Amsterdam, 1986.

[318] Pritchard D. E., E. L. Raab, V. Bagnato, C. E. Wieman, and R. N. Watts. Light traps using spontaneous forces. *Phys. Rev. Lett.*, 57: 310–313, 1986.

[319] Prodon J. V., W. D. Phillips, and H. J. Metcalf. Laser production of a very slow monoergetic atomic beam. *Phys. Rev. Lett.*, 49: 1149–1153, 1982.

[320] Grimm R., M. Weidemüller, and Y. B. Ovchinnikov. Optical dipole traps for neutral atoms. *Adv. At. Mol. Opt. Phys.*, 42: 95–113, 2000.

[321] Raab E., M. Prentiss, A. Cable, S. Chu, and D. E. Pritchard. Trapping of neutral sodium atoms with radiation pressure. *Phys. Rev. Lett.*, 57: 2632–2634, 1987.

[322] Rafac R. J., C. E. Tanner, A. E. Liningston, K. W. Kukla, H. G. Berry, and C. A. Kurtz. Precision lifetime measurements of the 6p $2P_{1/2,3/2}$ states in atomic cesium. *Phys. Rev. A*, 50: R1976–R1979, 1994.

[323] Ramirez-Serrano J., W. DeGraffenreid, and J. Weiner. Beam-loss spectroscopy of cold atoms in a bright sodium beam. *Phsy. Rev. A.* to be submitted.

[324] Ramirez-Serrano J., W. DeGraffenreid, and J. Weiner. Polarization-dependent spectra in the photoassociative ionization of cold atoms in a bright sodium beam. *Phys. Rev. A*, 65: 052719-1–052719-8, 2002.

[325] Ramsey N. F. *Molecular Beams*. Oxford University Press, London, 1956, reprinted 1963, 1969.

[326] Ratliff L. P., M. E. Wagshul, P. D. Lett, and S. L. Rolston, and W. D. Phillips. Photoassociative spectroscopy of $1_g$, $0_u^+$, and $0_g^-$ states of $Na_2$. *J. Chem Phys.*, 101: 2638–2641, 1994.

[327] Riis E., D. S. Weiss, K. A Moler, and S. Chu. Atom funnel for the production of a slow, high-density atomic beam. *Phys. Rev. Lett.*, 64: 1658–1661, 1990.

[328] Ritchie N. W. M., E. R. I. Abraham, and R. G. Hulet. Trap loss collisions of $^7Li$: the role of trap depth. *Laser Phys.*, 4: 1066–1075, 1994.

[329] Ritchie N. W. M., E. R. I. Abraham, Y. Y. Xiao, C. C. Bradley, R. G. Hulet, and P. S. Julienne. Trap-loss collisions of ultracold lithium atoms. *Phys. Rev. A*, 51: R890–R893, 1995.

[330] Robert A., A. Browaeys, J. Poupard, S. Nowak, D. Boiron, C. I. Westbrook, and A. Aspect. A Bose–Einstein condensate of metastable atoms. *Science*, 292: 461–464, 2001.

[331] Roberts J. L., J. P. Burke, N. R. Claussen, S. L. Cornish, E. A. Donley, and C. E. Wieman. Improved characterization of elastic scattering near a Feshbach resonance in $^{85}$Rb. *Phys. Rev. A*, 64: 024702-1–024702-3, 2001.

[332] Roberts J. L., N. R. Claussen, J. P. Burke, C. H. Greene, E. A. Cornell, and C. E. Wieman. Resonant magnetic field control of elastic scattering in cold $^{85}$Rb. *Phys. Rev. Lett.*, 81: 5109–5112, 1998.

[333] Roberts J. L., N. R. Claussen, S. L. Cornish, E. A. Donley, E. A. Cornell, and C. E. Wieman. Controlled collapse of a Bose–Einstein condensate. *Phys. Rev. Lett.*, 86: 4211–4214, 2001.

[334] Rolston S. NIST. Private communication, 1997.

[335] Sackett C. A., C. C. Bradley, and R. G Hulet. Optimization of evaporative cooling. *Phys. Rev. A*, 55: 3797–3801, 1997.

[336] Sanchez-Villicana V., S. D. Gensemer, and P. L. Gould. Observation of flux enhancement in collisions between ultracold atoms. *Phys. Rev. A*, 54: R3730–R3733, 1996.

[337] Sanchez-Villicana V. S., T. P. Dinneen, W. Süptitz, and P. L. Gould. Suppression of ultracold ground-state hyperfine-changing collision with laser light. *Phys. Rev. Lett.*, 74: 4619–4622, 1995.

[338] Santos L. and G. V. Shlyapnikov. Collapse dynamics of trapped Bose–Einstein condensates. *Phys. Rev. A*, 66: 011602-1–01602-4, 2002.

[339] Santos M. S., P. Nussenzveig, L. G. Marcassa, K. Helmerson, J. Flemming, S. C. Zilio, and V. S. Bagnato. Simultaneous trapping of two different atomic species in a vapor-cell magneto-optical trap. *Phys. Rev. A*, 52: R4340–R4343, 1995. See also erratum, *Phys. Rev. A* **54**, 1739 (1996).

[340] Santos M. S., P. Nussenzveig, S. Zilio, and V. S. Bagnato. Intensity dependence of the collisional loss rate for potassium atoms in a vapor cell. *Laser Phys.*, 8: 880, 1998.

[341] Scheingraber H. and C. R. Vidal. Discrete and continuous Franck Condon factors of the Mg$_2$ $A\,^1\Sigma_u^+-X\,^1\Sigma_g^+$ system and their J dependence. *J. Chem. Phys.*, 66: 3694–3704, 1977.

[342] Schlöder U., H. Engler, U. Schünemann, R. Grimm, and M. Weidmüller. Cold inelastic collisions between lithium and cesium in a two-species magneto-optical trap. *Eur. Phys. J. D*, 7: 331–340, 1999.

[343] Schreck F., K. L. Khaykovich, K. L. Corwin, G. Ferrari, T. Bourdel, J. Cubizolles, and C. Salomon. Quasipure Bose–Einstein condensate immersed in a Fermi sea. *Phys. Rev. Lett.*, 87: 080403-1–080403-4, 2001.

[344] Scoles G. *Atomic and Molecular Beam Methods*, volume I. Oxford University Press, New York, 1988.

[345] Sesko D., T. Walker, C. Monroe, A. Gallagher, and C. Wieman. Collisional losses from a light-force trap. *Phys. Rev. Lett.*, 63: 961–964, 1989.

[346] Sesko D. W., T. G. Walker, and C. Wieman. Behavior of neutral atoms in a spontaneous force trap. *J. Opt. Soc. Am. B*, 8: 946–958, 1991.

[347] Shaffer J. P., W. Chalupczak, and N. P. Bigelow. Ionization of heteronuclear molecules in a novel two species magneto-optical trap. *Phys. Rev. Lett.*, 82: 1124–1127, 1999.

[348] Shaffer J. P., W. Chalupczak, and N. P. Bigelow. Trap loss in a two-species Na-Cs magneto-optical trap: intramultiplet mixing in heteronuclear ultracold collisions. *Phys. Rev. A*, 60: R3365–R3368, 1999.

[349] Shang S.-Q., Z. T. Lu, and S. J. Freedman. Comparison of the cold-collision losses for laser-trapped sodium in different ground-state hyperfine sublevels. *Phys. Rev. A*, 50: R4449–R4452, 1994.

[350] Sheehy B., S.-Q. Shang, R. Watts, S. Hatamian, and H. Metcalf. Diode-laser deceleration and collimation of a rubidium beam. *J. Opt. Soc. Am. B*, 6: 2165–2170, 1989.

[351] Shlyapnikov G. V., J. T. M. Walraven, U. M. Rahmanov, and M. W. Reynolds. Decay kinetics and Bose condensation in a gas of spin-polarized triplet helium. *Phys. Rev. Lett.*, 73: 3247–3250, 1994.

[352] Shor P. W. Polynomial-time algorithms for prime factorization and discrete logarithms on a quantum computer. *arXiv e-print*, quantph/9508027, 1994.

[353] Silvera I. F. and J. T. M. Walraven. Stabilization of atomic hydrogen at low temperature. *Phys. Rev. Lett.*, 44: 164–168, 1980.

[354] Silvera I. F. and J. T. M. Walraven. *Spin-Polarized Atomic H*, volume X of *Progress in Low Temperature Physics*, pp. 139–370. Elsevier, Amsterdam, 1986. edited by D. F. Brewer.

[355] Smith A. M., K. Burnett, and P. S. Julienne. Semiclassical theory of collision-induced loss from optical traps. *Phys. Rev. A*, 46: 4091–4099, 1992.

[356] Söding J., D. Guery-Odelin, P. Desbiolles, G. Ferrari, M. Ben Dahan, and J. Dalibard. Giant spin relaxation of an ultracold cesium gas. *Phys. Rev. Lett.*, 80: 1869–1872, 1998.

[357] Solts R., A. Ben-Reuven, and P. S. Julienne. Optical collisions in ultracold atom traps: Two-photon distorted-wave theory. *Phys. Rev. A*, 52: 4029–4042, 1995.

[358] Sprik R., J. T. M. Walraven, G. H. Yperen, and I. F. Silvera. State-dependent recombination and suppressed nuclear relaxation in atomic hydrogen. *Phys. Rev. Lett.*, 49: 153–156, 1982.

[359] Stean A. Quantum computing. *arXiv e-print*, 2(quant-ph/9708022): 1–64, September 1997.

[360] Steane A., M. M. Chowdhury, and C. J. Foot. Radiation force in the magneto-optical trap. *J. Opt. Soc. Am. B*, 9: 2142–2158, 1992.

[361] Stenger J., S. Inouye, M. R. Andrews, H.-J. Miesner, D. M. Stamper-Kurn, and W. Ketterle. Strongly enhanced inelastic collisions in a Bose–Einstein condensate near feshbach resonances. *Phys. Rev. Lett.*, 82: 2422–2425, 1999.

[362] Stenholm. S. The semiclassical theory of laser cooling. *Rev. Mod. Phys.*, 58: 699–739, 1986.

[363] Stoof H. T. C., M. Bijlsma, and M. Houbiers. Theory of interacting quantum gases. *J. Res. Natl. Inst. Stand. Technol.*, 101: 443–457, 1996.

[364] Stoof H. T. C., A. M. L. Janssen, J. M. V. A. Koelman, and B. J. Verhaar. Decay of spin-polarized atomic hydrogen in the presence of a Bose condensate. *Phys. Rev. A*, 39: 3157–3169, 1989.

[365] Stoof H. T. C., J. M. V. A. Koelman, and B. J. Verhaar. Spin-exchange and dipole relaxation rates in atomic hydrogen: rigorous and simplified calculations. *Phys. Rev. B*, 38: 4688–4697, 1988.

[366] Stwalley W. C. The dissociation energy of the hydrogen molecule using long-range forces. *Chem. Phys. Lett.*, 6: 241–244, 1970.

[367] Stwalley W. C. and L. Nosanow. Possible new quantum systems. *Phys. Rev. Lett.*, 36: 910–913, 1976.

[368] Stwalley W. C., Y.-H. Uang, and G. Pichler. Pure long-range molecules. *Phys. Rev. Lett.*, 41: 1164–1166, 1978.

[369] Sukenik C. I., D. Hoffmann, S. Bali, and T. Walker. Low saturation intensities in two-photon ultracold collisions. *Phys. Rev. Lett.*, 81: 782–785, 1998.

[370] Suominen. K.-A. Theories for cold atomic collisions in light fields. *J. Phys. B*, 29: 5981–6007, 1996.

[371] Suominen K.-A., Y. B. Band, I. Tuvi, K. Burnett, and P. S. Julienne. Quantum and semiclassical calculations of cold collisions in light fields. *Phys. Rev. A*, 57: 3724–3738, 1998.

[372] Suominen K.-A., K. Burnett, and P. S. Julienne. Role of off-resonant excitation in cold collisions in a strong laser field. *Phys. Rev. A*, 53: R1220–R1223, 1996.

[373] Suominen K.-A., K. Burnett, P. S. Julienne, M. Walhout, U. Sterr, C. Orzel, M. Hoogerland, and S. L. Rolston. Ultracold collisions and optical shielding in metastable xenon. *Phys. Rev. A*, 53: 1658–1689, 1996.

[374] Suominen K.-A., M. J. Holland, K. Burnett, and P. S. Julienne. Optical shielding of cold collisions. *Phys. Rev. A*, 51: 1446–1457, 1995.

[375] Suominen K.-A., E. Tiesinga, and P. S. Julienne.

[376] Suominen K.-A., M. J. Holland K. Burnett, and P. S. Julienne. Excited-state survival probabilities for cold collisions in a weak laser field. *Phys. Rev. A*, 49: 3897–3902, 1994.

[377] Swanson T. B., N. J. Silva, S. K. Mayer, J. J. Maki, and D. H. McIntyre. Rubidium atomic funnel. *J. Opt. Soc. Am. B*, 13: 1833–1836, 1996.

[378] Takekoshi T., B. M. Patterson, and R. J. Knize. Observation of optically trapped cold cesium molecules. *Phys. Rev. Lett.*, 81: 5105–5108, 1998.

[379] Telles G. D., L. S. Aguia, L. G. Marcassa, and V. S. Bagnato. Evidence of a two-color trap-loss channel. *Phys. Rev. A*, 66: 025403-1–025403-4, 2002.

[380] Telles G. D., V. S. Bagnato, and L. G. Marcassa. Alternative to the hyperfine-change-collision interpretation for the behavior of magneto-optical-trap losses at low light intensity. *Phys. Rev. Lett.*, 86: 4496–4499, 2001.

[381] Telles G. D., W. Garcia, L. G. Marcassa, V. S. Bagnato, D. Ciampini, M. Fazzi, J. H. Müller, D. Wilkowski, and E. Arimondo. Trap loss in a two-species Rb–Cs magneto-optical trap. *Phys. Rev. A*, 63: 033406-1–033406-7, 2001.

[382] Telles G. D., L. G. Marcassa, S. R. Muniz, S. G. Miranda, A. Antunes, C. Westbrook, and V. S. Bagnato. Inelastic cold collisions of a Na/Rb mixture in a magneto-optical trap. *Phys. Rev. A*, 59: R23–R26, 1999.

[383] Tellinghuisen J. *Photodissociation and Photoionization*, volume LX of *Advances in Chemical Physics*, pp. 299–369. Wiley, New York, 1985.

[384] Thorsheim H. R., Y. Wang, and J. Weiner. Cold collisions in an atomic beam. *Phys. Rev. A*, 41: 2873–2876, 1990.

[385] Thorsheim H. R., J. Weiner, and P. S. Julienne. Laser-induced photoassociation of ultracold sodium atoms. *Phys. Rev. Lett.*, 58: 2420–2423, 1987.

[386] Tiesinga E., S. B. Crampton, B. J. Verhaar, and H. T. C. Stoof. Collisional frequency shifts and line broadening in the cryogenic deuterium maser. *Phys. Rev. A*, 47: 4342–4347, 1993.

[387] Tiesinga E., S. Kotochigova, and P. S. Julienne. Scattering length of the ground-state Mg+Mg.

[388] Tiesinga E., S. J. M. Kuppens, B. J. Verhaar, and H. T. C. Stoof. Collisions between cold grond state Na atoms. *Phys. Rev. A*, 43: R5188–R5191, 1991.

[389] Tiesinga E., A. J. Moerdijk, B. J. Verhaar, and H. T. C. Stoof. Conditions for Bose–Einstein condensation in magnetically trapped atomic cesium. *Phys. Rev. A*, 46: R1167–R1170, 1992.

[390] Tiesinga E., H. T. C. Stoof, and B. J. Verhaar. Spin-exchange frequency shift of the cryogenic deuterium maser. *Physica B*, 165: 19–20, 1990.

[391] Tiesinga E., B. J. Verhaar, and H. T. C. Stoof. Threshold and resonance phenomena in ultracold ground-state collisions. *Phys. Rev. A*, 47: 4114–4122, 1993.

[392] Tiesinga E., B. J. Verhaar, H. T. C. Stoof, and D. van Bragt. Spin-exchange frequency shift in cesium atomic fountain. *Phys. Rev. A*, 45: 2671–2674, 1992.

[393] Tiesinga E., C. J. Williams, and P. S. Julienne. Photoassociative spectroscopy of highly excited vibrational levels of alkali dimers: Green's function approach for eigenvalue solvers. *Phys. Rev. A*, 57: 4257–4267, 1998.

[394] Tiesinga E., C. J. Williams, P. S. Julienne, K. M. Jones, P. D. Lett, and W. D. Phillips. A spectroscopic determination of scattering lengths for sodium atom collisions. *J. Res. Natl. Inst. Stand. Technol.*, 101: 505–520, 1996.

[395] Timmermans E., P. Tommasini, R. Côté, M. Hussein, and A. Kerman. Inter-condensate tunneling in Bose–Einstein condensates with Feshbach resonances. *e-print arXiv*, 1(cond-mat/9805323): 1–11, May 1998.

[396] Toennies J. P. and A. Vilesov. Spectroscopy of atoms and molecules in liquid helium. *Ann. Rev. Phys. Chem.*, 49: 1–41, 1998.

[397] Tommasini P., E. Timmermans, M. Hussein, and A. Kerman. Feshbach resonance and hybrid atomic/molecular BEC-systems. *e-print arXiv*, 1(cond-mat/9804015): 1–11, April 1998.

[398] Townsend C., W. Ketterle, and S. Stringari. Bose–Einstein condensation. *Phys. World*, 10: 29–34, 1997.

[399] Treutlein P., K. Y. Chung, and S. Chu. High-brightness atom source for atomic fountains. *Phys. Rev. A*, 63: 051401-1–051401-4, 2001.

[400] Trost J., C. Eltschka, and H. Friedrich. Quantization in molecular potentials. *J. Phys. B*, 31: 361–374, 1998.

[401] Tsai C. C., R. S. Freeland, J. M Vogels, H. M. J. M. Boesten, B. J. Verhaar, and D. J. Heinzen. Two-color photoassociation spectroscopy of ground state Rb$_2$. *Phys. Rev. Lett.*, 79: 1245–1248, 1997.

[402] Tsao C.-C. *Photoassociative Ionization of Cold Sodium Atoms in an Atomic Beam*. PhD thesis, University of Maryland, 1996.

[403] Tsao C.-C., R. Napolitano, Y. Wang, and J. Weiner. Ultracold photoassociative ionization collisions in an atomic beam: optical field intensity and polarization dependence of the rate constant. *Phys. Rev. A*, 51: R18–R21, 1995.

[404] Tsao C.-C., W. Wang, J. Weiner, and V. S. Bagnato. Optical collimation and compression of a thermal atomic beam. *J. Appl. Phys.*, 80: 8–14, 1996.

[405] Tsao C.-C., Y. Wang, R. Napolitano, and J. Weiner. Anisotropy of optical suppression in phtoassociative ionization collisions within a slow, collimated sodium atom beam. *Eur. Phys. D*, 4: 139–144, 1998.

[406] Uang Y.-H. and W. C. Stwalley. Close-coupling calculations of spin-polarized hydrogen–deuterium collisions. *Phys. Rev. Lett.*, 45: 627–530, 1980.

[407] Ungar P. J., D. S. Weiss, E. Riis, and S. Chu. Optical molasses and multilevel atoms: theory. *J. Opt. Soc. Am. B*, 6: 2058–2071, 1989.

[408] van Abeelen F. A. and B. J. Verhaar. Determination of collisional properties of cold Na atoms from analysis of bound-state photoassociation and Fesbhach resonance field data. *Phys. Rev. A*, 59: 578–584, 1999.

[409] van Abeelen F. A. and B. J. Verhaar. Time-dependent Feshbach resonance scattering and anomalous decay of a Na Bose–Einstein Condensate. *Phys. Rev. Lett.*, 83: 1550–1553, 1999.

[410] van Roijen R., J. J. Berkhout, S. Jaakkola, and J. T. M. Walraven. Experiments with atomic hydrogen in a magnetic trapping field. *Phys. Rev. Lett.*, 61: 931–934, 1988.

[411] Vardi A., D. Abrashkevich, E. Frishman, and M. Shapiro. Theory of radiative recombination with strong laser pulses and the formation of ultracold molecules via stimulated photo-recombination of cold atoms. *J. Chem. Phys.*, 107: 6166–6174, 1997.

[412] Vardi A., V. A Yurovsky, and J. R. Anglin. Quantum effects on the dynamics of a two-mode atom-molecule Bose–Einstein condensate. *Phys. Rev. A*, 64: 063611-1–063611-5, 2001.

[413] Verhaar B. J., K. Gibble, and S. Chu. Cold-collision properties derived from frequency shifts in a Cs fountain. *Phys. Rev. A*, 48: R3429–R3432, 1993.

[414] Verhaar B. J., J. M. V. A. Koelman, H. T. C. Stoof, O. J. Luiten, and S. B. Crampton. Hyperfine contribution to spin-exchange frequency shifts in the hydrogen maser. *Phys. Rev. A*, 35: 3825–3831, 1987.

[415] Vigué. J. Possibility of applying laser-cooling techniques to the observation of collective quantum effects. *Phys. Rev. A*, 34: 4476–4479, 1986.

[416] Vogels J. M., C. C. Tsai, R. S. Freeland, S. J. J. M. F. Kokkelmans, B. J. Verhaar, and D. J. Prediction of Feshbach resonances in collisions of ultracold rubidium atoms. *Phys. Rev. A*, 56: R1067–R1070, 1997.

[417] Wagshul M. E., K. Helmerson, P. D. Lett, L. Rolston, W. D. Phillips, R. Heather, and P. S. Julienne. Hyperfine effects on associative ionizatioin of ultracold sodium. *Phys. Rev. Lett.*, 70: 2074–2077, 1993.

[418] Walhout M., U. Sterr, C. Orzel, M. Hoogerland, and S. L. Rolston. Optical control of ultracold collisions in metastable xenon. *Phys. Rev. Lett.*, 74: 506–509, 1995.

[419] Walker T. and P. Feng. Measurements of collisions between laser-cooled atoms. *Adv. At. Mol. Opt. Phys.*, 34: 125–170, 1994.

[420] Walker T. and D. R. Pritchard. Effects of hyperfine structure on alkali trap-loss collisions. *Laser Phys.*, 4: 1085–1092, 1994.

[421] Walker T., D. Sesko, and C. Wieman. Collective behavior of optically trapped neutral atoms. *Phys. Rev. Lett.*, 64: 408–411, 1990.

[422] Wallace C., T. Dinneen, K. Tan, T. Grove, and P. Gould. Isotopic difference in trap loss collisions of laser cooled rubidium atoms. *Phys. Rev. Lett.*, 69: 897–900, 1992.

[423] Wallace C., V. Sanchez-Villicana, T. P. Dinneen, and P. L. Gould. Suppression of trap loss collisions at low temperature. *Phys. Rev. Lett.*, 74: 1087–1090, 1995.

[424] Walraven J. T. M. *Atomic Hydrogen at Sub-Kelvin Temperatures*, volume 9, pp. 187–211. World Scientific, Singapore, 1984. edited by R. S. Dyck, Jr. and E. N. Fortson.

[425] Walsworth R. L., I. F. Silvera, H. P. Gotfried, C. C. Agosta, R. F. C. Vessot, and E. M. Mattison. Hydrogen maser at temperatures below 1 K. *Phys. Rev. A*, 34: 2550–2553, 1986.

[426] Wang H., P. L. Gould, and W. C. Stwalley. Photoassociative spectroscopy of pure long-range molecules. *Z. Phys. D*, 36: 317–323, 1996.

[427] Wang H., P. L. Gould, and W. C. Stwalley. Photoassociative spectroscopy of ultracold $^{39}$K atoms in a highg-density vapor-cell magneto-optical trap. *Phys. Rev. A*, 53: R1216–R1219, 1996.

[428] Wang H., P. L. Gould, and W. C. Stwalley. Long-range interaction of the $^{39}$K($4s$) + $^{39}$K($4p$) asymptote by photoassociative spectroscopy. I: The $0_g^-$ pure long-range state and the long-range potential constants. *J. Chem. Phys.*, 106: 7899–7912, 1997.

[429] Wang H., P. L. Gould, and W. C. Stwalley. Fine structure predissociation of ultracold photoassociated $^{39}$K$_2$ molecules observed by fragmentation sptectroscopy. *Phys. Rev. Lett.*, 80: 476–479, 1998.

[430] Wang H., J. Li, X. T. Wang, C. J. Williams, P. L. Gould, and W. C. Stwalley. Precise determination of the dipole matrix element and radiative lifetime of the $^{39}$K $4p$ state by photoassociative spectroscopy. *Phys. Rev. A*, 55: R1569–R1572, 1997.

[431] Wang H., A. N. Nikolov, J. R. Ensher, P. L. Gould, E. E. Eyler, W. C. STwalley, J. P. Burke, J. L. Bohn, C. H. Greene, E. Tiesinga, C. J. Williams, and P. S. Julienne. Ground-state scattering lengths for potassium isotopes determined by double-resonance photoassociative spectroscopy of ultracold $^{39}$K. *Phys. Rev. A*, 62: 052704-1–052704-4, 2000.

[432] Wang H. and W. C. Stwalley. Ultracold photoassociation spectroscopy of heteronuclear alkali-metal diatomic molecules. *J. Chem. Phys.*, 108: 5767–5771, 1998.

[433] Wang H., X. T. Wang, P. L. Gould, and W. C. Stwalley. Optical–optical double resonance photoassociative spectroscopy of ultracold $^{39}$K atoms near highly excited asymptotes. *Phys. Rev. Lett.*, 78: 4173–4176, 1997.

[434] Wang M.-X., J. Keller, J. Boulmer, and J. Weiner. Strong velocity dependence of the atomic alignment effect in Na(3p) + Na(3p) associaitve ionization. *Phys. Rev. A*, 34: 4497–4500, 1986.

[435] Wang M.-X., J. Keller, J. Boulmer and J. Weiner. Spin-selected velocity dependence of the associative ionization cross section in Na(3p) + Na(3p) associative ionization ovet the collision energy range from 2.4 to 290 meV. *Phys. Rev. A*, 35: 934–937, 1987.

[436] Wang X., H. Wang, P. L. Gould, W. C. Stwalley, E. Tiesinga, and P. S. Julienne. Observation of the pure long range $1_u$ state of an alkali-metal dimer in photoassociation spectrosocpy. *Phys. Rev. A*, 57: 4600–4603, 1998.

[437] Wang Y. and J. Weiner. Velocity-selected atomic-beam collisions in the energy range from 300 to 5 K. *Phys. Rev. A*, 42: 675–677, 1990.

[438] Weber T., J. Herbig, M. Mark, H.-C. Nägerl, and R. Grimm. Bose–Einstein Condensation of Cesium. *Sciencexpress/www.Sciencexpress.org*, pp. 1–4, 5 December 2002.

[439] Weiner J. Experiments in cold and ultracold collisions. *J. Opt. Soc. Am. B*, 6: 2270–2278, 1989.

[440] Weiner J. Advances in ultracold collisions: experimentation and theory. *Adv. At. Mol. Opt. Phys.*, 35: 45–78, 1995.

[441] Weiner J., V. S. Bagnato, S. Zilio, and P. S. Julienne. Experiments and theory in cold and ultracold collisions. *Rev. Mod. Phys.*, 71: 1–85, 1999.

[442] Weiner J., F. Masnou-Seeuws, and A. Giusti-Suzor. Associative ionization: experiments, potentials and dynamics. *Adv. At. Mol. Opt. Phys.*, 26: 209–296, 1989.

[443] Weinstein J. D., R. deCarvalho, T. Guillet, B. Friedrich, and J. M. Doyle. Magnetic trapping of calcium monohydride molecules at millikelvin temperatures. *Nature*, 395: 148–150, 1998.

[444] Wigner E. P. On the behavior of cross sections near thresholds. *Phys. Rev.*, 73: 1002–1009, 1948.

[445] Williams C. J. and P. S. Julienne. Mass effects in the theoretical determination of nuclear spin relaxation rates for atomic hydrogen and deuterium. *Phys. Rev. A*, 47: 1524–1527, 1993.

[446] Williams C. J. and P. S. Julienne. Molecular hyperfine structure in the photoassociation spectroscopy of laser cooled atoms. *J. Chem. Phys.*, 101: 2634–2637, 1994.

[447] Williams C. J., E. Tiesinga, and P. S. Julienne. Hyperfine structure of the Na$_2$ $0_g^-$ long-range molecular state. *Phys. Rev. A*, 53: R1939–R1942, 1996.

[448] Williamson R. S. and T. Walker. Magneto-optical trapping and ultracold collisions of potassium atoms. *J. Opt. Soc. Am. B*, 12: 1393–1397, 1995.

[449] Wineland D. J. and W. M. Itano. Laser cooling of atoms. *Phys. Rev. A*, 20: 1521–1540, 1979.

[450] Wynar R., R. S. Freeland, D. J. Han, C. Ryu, and D. J. Heinzen. Molecules in a Bose–Einstein condensate. *Science*, 287: 1016–1019, 2000.

[451] Yu J., J. Djemaa, P. Nosbaum, and P. Pillet. Funnel with orientated Cs atoms. *Opt. Commun.*, 112: 136–140, 1994.

[452] Yurovsky V. A. and A. Ben-Reuven. Incomplete optical shielding in cold atom traps: three-dimensional Landau–Zener theory. *Phys. Rev. A*, 55: 3772–3779, 1997.

[453] Zilio S. C., L. Marcassa, S. Muniz, R. Horowicz, V. Bagnato, R. Napolitano, J. Weiner, and Julienne P. S. Polarization dependence of optical suppression in photoassociative ionization collisions in a sodium magneto-optic trap. *Phys. Rev. Lett.*, 76: 2033–2036, 1996.

# Index

atom beams, ix, 37, 122

BEC, 2, 129, 155, 156, 157, 159, 176
Bose-Einstein condensation, *see* BEC
bosons, 5, 12, 157

cold molecules, 111, 179
   and stimulated processes, 183
Condon point, 3, 24, 25, 47, 101, 104, 129
cross section, 5, 7–9, 11, 13, 20, 27, 43, 50, 103

dipole trap, 102, 104, 105, 176, 177, 181
Doppler broadening, 5
Doppler cooling, 1
   and sub-Doppler cooling, 92

Earnshaw theorem, 31
elastic scattering, 5, 9, 12, 13, 17, 41, 159, 165, 172
evaporative cooling, ix, 1, 5, 12, 13, 41, 136, 155, 157–159, 166, 167, 172, 173, 176, 177, 179, 194

far-off resonance trap, *see* FORT
FCC, 41, 42, 45, 46, 64, 66–68, 73, 88
fermions, 12
Feshbach resonances, 119, 133, 158, 159, 165, 169, 169, 170, 176
   and molecular BEC, 186
fine-structure-changing collision, *see* FCC
FORT, 35, 97, 112, 117, 134, 174
Franck-Condon factor, 24

Gaussian, 58, 60, 66, 82
GOST trap, 173, 174, 177

HCC, 41, 42, 66, 68, 69, 80, 89, 89, 90, 140, 141
   cesium and rubidium, 90
   low-intensity, 94
   sodium, 90
hyperfine-changing collisions, *see* HCC

Landau-Zener, 44, 49, 50, 53–55, 138, 142, 145, 152
line shape, 47, 50, 107, 116, 135
   Lorentzian, 30, 47, 55, 107

magneto-optical trap, *see* MOT
MOT, ix, 31, 34, 41, 42, 44, 59, 60, 63, 64, 67, 69, 76, 77, 79, 83, 87, 93, 100, 117, 121, 141
   dark spot, 112, 113
   dual, 194

optical molasses, 1, 102–104, 106
optical shielding, 4, 25, 138, 146, 148, 149
optical suppression, 39, 139, 140, 142, 143, 145, 148, 152

PAI, 40, 83, 99, 101, 104, 106, 111, 112
   potassium
      scattering length, 135
   rubidium
      two-color, 119
   shielding and suppression, 142
   sodium, 114
      atom beams, 122
      scattering length, 131
      two-color, 117
partial waves, 4, 11, 12, 17, 18, 26, 40, 51, 88, 103, 114, 116, 128, 147, 153
phase shift, 2, 8, 13, 15, 17, 19–22, 95, 129, 130, 136, 156, 164, 191
photoassociation, ix, 3, 4, 25, 26, 50, 52, 56, 84, 97
   ambient and cold temperatures, 99
   and BEC, 167
   and cold molecules, 179, 182
   and Feshbach resonances, 165
   atomic lifetimes, 125
      table, 128
   lithium, 120
   potassium, 121
   rubidium, 117
      two-color, 119